D0812277

Kansas Geology

Kansas Geology

An Introduction to Landscapes, Rocks, Minerals, and Fossils

EDITED BY REX BUCHANAN

PUBLISHED FOR THE
KANSAS GEOLOGICAL SURVEY BY THE

UNIVERSITY PRESS OF KANSAS

Landscapes, except where marred by some human maltreatment, show perfect harmony in all their parts, and I submit Kansas, without lofty mountains or awe-inspiring canyons, as a thing of beauty composed of placid forms. It was created less violently than some other parts of the world but by forces just as relentless and just as exciting.

—John Mark Jewett

© 1984 by the University Press of Kansas
All rights reserved
Second printing, 1985
Third printing, 1989

Published by the University Press of Kansas (Lawrence, Kansas 66045),
which was organized by the Kansas Board of Regents and is operated and funded
by Emporia State University, Fort Hays State University, Kansas State
University, Pittsburg State University, the University of Kansas,
and Wichita State University

Library of Congress Cataloging in Publication Data

Main entry under title:

Kansas geology.

Includes index.
1. Geology—Kansas. I. Buchanan, Rex, 1953-
II. Kansas Geological Survey.
QE113.K3   1983   557.81   83-23546
ISBN 0-7006-0239-9
ISBN 0-7006-0240-2 (pbk.)

Printed in the United States of America
Designed by Larry Hirst

Contents

List of Color Illustrations

Between pages 52–53

Preface

by Rex Buchanan

*The landscape . . . either developed
or wild, is an historical document.*
                    —Roderick Nash

In a square foot of Kansas soil, a friend of mine once proclaimed, more history is recorded than in all of the history books in the state's libraries. Although perhaps guilty of overstatement, he had a point. The rocks and soils of Kansas contain all sorts of stories—of seas inhabited by ferocious sharks, of swamps that stretched for miles, and of dust storms that make the 1930s Dust Bowl look clean by comparison. Ancient oceans, saline seas larger than the Great Salt Lake, earthquakes, and even volcanoes are recorded in Kansas rocks if you know how to read them.

But reading rocks is not as easy as opening a book. Much of the state's geologic history is buried beneath the surface. Extracting that history, then, requires drilling or mining for samples or the use of geophysics to collect indirect information about subterranean rock formations. Reading even surface rocks demands specialized knowledge. Take the huge blocks of limestone that motorists see lining the highways of eastern Kansas. Geologists know that limestone is deposited at the bottom of seas, far from shore, where the chemical composition of the water is exactly right. Slight changes in the water's composition cause the limestone to take on different colors and characteristics. In one part of the sea red-streaked limestone might be produced; in another, it could become more buff-colored. Limestone found in one area might make building stone and fenceposts; in another, its only use may be as gravel.

This book is intended not for those who can already read limestone and other rocks but rather for those who have little or no background in earth science but want to learn more about the geology of Kansas. An effort has been made to present the required technical information in a way that avoids distorting oversimplification on the one hand and mystifying complexity on the other.

An introduction briefly summarizes several complex topics. It attempts to sketch Kansas geologic history; how Kansas fits into the geology of the North American continent; the concepts of geologic time and the labeling of rock formations; and the natural forces of deposition and erosion. The first chapter, by Frank Wilson, then describes the landscape, the rock layers that are visible at the surface in Kansas, grouping them according to the ages in which they were deposited. Hence, it begins with southeastern Kansas, where the oldest surface rocks are found, and works across the Kansas landscape to Colorado. The next two chapters, by Laura Tolsted and Ada Swineford, describe in some detail the common rocks and minerals found in Kansas: what they are composed of, how they were formed, where they are found, and how they can be identified. While most of the rocks discussed are sedimentary, others—whether native or transported into the state—are identified as igneous or metamorphic in origin. The fourth chapter, by Debra K. Bennett, describes the fossils and the life-forms that preceded them, explaining why Kansas is world-famous among fossil collectors. In the course of recounting the state's geologic history, she discusses the plants and animals of each period and paints a picture of the landscape and geology that prevailed. The final section returns the reader to the present by providing a guide to the geology visible to the motorist who crosses the state on U.S. Interstate 70. In addition to locating many of the geologic features discussed in earlier chapters, the guide gives additional information about Kansas geology and history. Finally, the reader will find a glossary, selected bibliography, and index at the back. For the reader who desires to go beyond this introduction to Kansas geology, the list of books recommended for additional reading should prove especially helpful.

Three chapters of this book are based on publications previously issued by the Kansas Geological Survey. The chapter on landscapes is derived from a longer booklet published in 1978 in the Survey's Educational Series. The second and third chapters are taken from the most popular title in the same series, *Kansas Rocks and Minerals,* which first appeared in 1948 but has been reprinted over a dozen times. All three chapters reflect revision—and, in some cases, updating—for this book. The remainder of the text appears here for the first time. All photographs, both in color and black and white, were taken expressly for this book; also new are the three color drawings (Plates 29, 30, and 31), as well as about half of the fossil drawings. The chapter on fossils was drafted especially for this book, but some of the material is drawn from a work in progress, *Vertebrate Fossils of Kansas,* to be published jointly by the University of Kansas Museum of Natural History and the Kansas Geological Survey.

The Kansas Geological Survey is a research and service division of the University of Kansas, and was created in 1889 to study the state's natural resources. During that time the Survey has produced a multitude of publications that describe the state's geology. Many of those reports and maps are technical, of interest primarily to earth scientists who regularly study the state's geology. But the Survey also publishes nontechnical, educational materials that are designed to help students and general readers understand Kansas geology. A list of those technical and nontechnical publications is available from the Survey upon request.

A number of people have contributed to the making of this book. George Barberich, Pieter Berendsen, Larry Brady, Nancy Christensen, Renate Hensiek, William Heard, Stan Roth, and Don Steeples contributed graphics and read various phases of the manuscript. John Charlton, Jim McCauley, Frank Wilson, and I took most of the photographs that appear in the book. William Hambleton, the director of the Survey, and Dean Lebestky, the Survey's associate director, provided invaluable administrative support that did much to improve the book's content and appearance.

Special acknowledgment must be made for the chapter on Kansas fossils. Philip S. Humphrey gave permission for reproduction of figures intended for publication in *Vertebrate Fossils of Kansas.* E. R. Hall gave permission for reproduction of some figures that have appeared in his book *The Mammals of North America,* 2d ed. (New York: John Wiley and Sons, 1981). John D. Chorn, J. D. Stewart, and L. D. Martin provided references and help in preparing the manuscript. Research for fossil fish reconstruction was carried out under the direction of John Chorn and J. D. Stewart. Randall E. Moss provided references and advice on coloration of fossil fishes. Research for amphibian and reptile reconstructions was reviewed by H.-P. Schultze. Several of the photographs are by John Chorn, Kenneth Whetstone, and Robert Reisz. Except where noted, the drawings are by Debra Bennett.

Introduction

by Rex Buchanan

Traveling in Kansas, it's easy to get the impression that the state's geology is simple and straightforward. First there's the landscape. While the state claims the Flint Hills and the chalk monuments of western Kansas, it boasts none of the towering mountain ranges or volcanoes found farther west. Even when drivers take the time to notice rock formations along their way, they may only see common, ordinary rocks such as limestone or sandstone. No lava, no mountains. Simple.

But things are not always what they seem. The geology of Kansas reveals all sorts of stories about the past, when we take the time to notice. A typical trek across the state takes drivers over territory that was—at one time or another—a shallow sea, a dismal swamp, or a vast salt plain. For example, the limestone of eastern Kansas shows that today's drivers are crossing an area that was once the floor of an ancient bay. Locked within the limestone are fossils of snails and clams and other life, a sort of permanent census of the population at the sea's bottom.

A little farther west, in Riley County, travelers pass over a series of faults that once produced an earthquake so strong that it shook cities as far away as northeastern Iowa. Those faults are still active, still producing small tremors that are occasionally felt by the residents above. North of Manhattan are a series of volcanolike domes that popped to the surface about 100 million years ago; in places, the craters of those eruptions are still visible.

As a traveler drives into the Flint Hills, bands of chert appear in the limestone, showing that the chemical composition of the sea must have changed slightly when these rocks were deposited. Even farther west, the roads pass through hills that are capped by red sandstone, rocks that were deposited near the shore of the old oceans, much like sand along a beach. Farther on, the limestone changes to chalk, and the fossils here are the remains of fierce animals—huge sharks and swimming reptiles—that are a little more frightening than the inhabitants of those older, more sedate seas of eastern Kansas.

Many of these features—like the Flint Hills or the chalk formations—are well known to the state's residents and travelers. But other geologic features are just as dramatic, although less familiar.* In south-central Kansas, the Red Hills form a desertlike landscape in Barber, Comanche, and Clark counties. In the northeastern corner of the state, glaciers have visibly rearranged the landscape and left behind huge boulders transported from hundreds of miles away. Scattered across the state, particularly in the areas underlain by salt, are sinkholes, some more than a mile across. Springs, caves, and waterfalls are also tucked away, hidden. The result is a sort of subtle diversity that says much about the geology of Kansas.

In short, the history of Kansas geology is not as simple as it might seem. The state's geology has produced a landscape that is more complex than most visitors, or many residents, realize. To understand Kansas geology, it may be best to begin with several generalizations.

First, nearly all the rocks visible on the state's surface are of one type: sedimentary. Over eons of time, sediments were deposited by rivers or winds or oceans and then compressed into layers of rock; they include the sandstones, shales, and limestones that are all commonly found in Kansas. Although igneous or metamorphic rocks appear at the surface in a few locations—those volcanolike features in Riley County represent one of the few outcrops of igneous rock in Kansas—nearly all rocks at the surface were created from sediments.

Second, almost without fail, the rocks in Kansas get older as you go deeper. During the state's geologic history, one layer of rock was deposited, then it was covered by a younger layer in a process that continued over hundreds of millions of years. Oceans covered Kansas and deposited, say, a layer of limestone, and then receded and deposited a layer of sandstone or shale. The effect is much like a giant layer cake. One flat layer of rock lies on top of another, with the older rocks underneath and the younger rocks on top. In some ways, that arrangement makes Kansas geology easy to decipher, easier to understand than in other parts of the world where layers of rock have been raised up or folded to create mountains, or where volcanoes and earthquakes have visibly rearranged the landscape.

In Kansas, variation in the geology is less striking. The state lies about in the middle of the North American continent, far away from the stormy edges of the continental plate where volcanoes and earthquakes abound. Kansas is in a much quieter zone that geologists call the central stable region. To the west are the Rocky Mountains, and to the east are the Appalachians, an older mountain range that was created in much the same

---

*Plates 1–32, found between pages 52 and 53, illustrate in color the diversity of Kansas' geology. Figures 1–121, which are black-and-white illustrations, are placed appropriately in various chapters to provide graphic accompaniment to the text, ranging from views of geologic features to drawings of fossils and prehistoric creatures.

ERAS	PERIODS	EPOCHS	EST. LENGTH IN YEARS	TYPE OF ROCK IN KANSAS	MILLION YEARS PAST
CENOZOIC	QUATERNARY	HOLOCENE 	10,000 +	Glacial drift; river silt, sand, and gravel; dune sand; wind-blown silt (loess); volcanic ash.	0.010
		PLEISTOCENE	1,990,000		2
	TERTIARY	PLIOCENE	3,000,000	River silt, sand, gravel, fresh-water limestone; volcanic ash; bentonite; diatomaceous marl; opaline sandstone.	5
		MIOCENE	19,000,000		24
		OLIGOCENE 	14,000,000		38
		EOCENE	17,000,000		55
		PALEOCENE	8,000,000		63
MESOZOIC	CRETACEOUS 		75,000,000	Limestone, chalk, chalky shale, dark shale, varicolored clay, sandstone, conglomerate. Outcropping igneous rock.	138
	JURASSIC		67,000,000	Sandstones and shales, chiefly subsurface. Siltstone, chert, and gypsum.	205
	TRIASSIC		35,000,000		240
PALEOZOIC	PERMIAN 		50,000,000	Limestone, shale, evaporites (salt, gypsum, anhydrite), red sandstone; chert, siltstone, dolomite, and red beds.	290
	PENNSYLVANIAN 		40,000,000	Alternating marine and nonmarine shale, limestone, sandstone, coal; chert and conglomerate.	~330
	MISSISSIPPIAN		30,000,000	Limestone, shale, dolomite, chert, oölites, sandstone, and siltstone.	360
	DEVONIAN 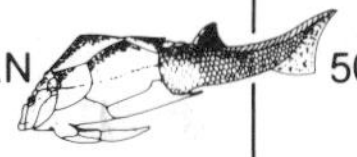		50,000,000	Subsurface only. Limestone, predominantly black shale; sandstone.	410
	SILURIAN 		25,000,000	Subsurface only. Limestone.	435
	ORDOVICIAN		65,000,000	Subsurface only. Dolomite, sandstone.	500
	CAMBRIAN 		70,000,000	Subsurface only. Dolomite, sandstone, limestone, and shale.	~570
PRECAMBRIAN			1,930,000,000	Subsurface only. Granite, other igneous rocks, and metamorphic rocks.	
			1,100,000,000 +		2,500

Figure 1—The chart shows the approximate duration of geologic periods in Kansas, along with examples of the rocks deposited and animals that lived during these times. The rocks are successively younger as the column rises.

way as the Rockies. This kind of deformation was rare in the middle of North America, where the area sits atop a steady platform of igneous and metamorphic rocks. During the past 400 million years a veneer of sediments has been deposited, layer after layer, on top of that foundation. The area has been subjected to relatively little faulting or folding.

The surface of Kansas reflects that quiet geologic history. The eastern half of the state is part of a physiographic province that is labeled the Central Lowlands, a level area of slightly rolling hills sitting atop sedimentary rocks. This lowlands stretches from Texas to northern Minnesota and includes much of the country around the Great Lakes. The western half of Kansas is in the Great Plains province, a series of river-deposited sediments, which rise up to about six thousand feet in altitude before meeting the Rocky Mountains to the west.

Still, exploring Kansas geology is a complex task, in part because it requires the use of so many concepts that are foreign to everyday thinking. For example, understanding geology requires thinking in terms of long periods of time, huge sweeps of time that are hard to comprehend. For most of us, a decade or even a year is a long period of time; in geologic history, a year is hardly the blink of an eye. Some processes, such as erosion, may be barely noticeable over the course of a year; but during millions of years, erosion can have a dramatic, devastating effect on the landscape. As a result, geologists constantly think of much grander scales of time, millions and millions of years. For the average person, it is difficult to picture central Kansas covered by a saline sea for thousands of years; for geologists, it's part of the job.

Central to geology, then, is the concept of geologic time. Because the span of time since the earth was created is so great, geologists illustrate it by using a scale that is easier to comprehend. One such analogy compares the 4.5 billion years of earth history with a single calendar year. On that scale the oldest rocks that have so far been discovered at the earth's surface would date from mid-March. Primitive living things first appeared in the Precambrian seas in late November, and the coal-forming swamps of the Pennsylvanian period flourished for about four days in early December. Dinosaurs were the dominant animals in mid-December, but disappeared the day after Christmas, at about the same time that the Rocky Mountains were being pushed up. Manlike creatures appeared during the evening of December 31, and Columbus discovered America at three seconds before midnight. All in all, the geologic history of Kansas encompasses a time span that is nearly beyond our comprehension.

To distinguish between different times during geologic history, geologists have produced a variety of labels for periods of the past. The oldest and longest of the geologic periods is known as the Precambrian, and it includes the time from the creation of the earth up to about 600 million years ago (these periods are shown in Figure 1). Most of the igneous and metamorphic rock that underlies Kansas is Precambrian in age. Following the Precambrian were ages that geologists have labeled the Cambrian, Ordovician,

Silurian, and Devonian. The oldest rock that appears at the surface in Kansas, found in the extreme southeastern corner of the state, was deposited during the Mississippian period, about 325 million years ago. Rocks deposited during the next period, the Pennsylvanian, are common throughout the eastern third of Kansas and include the coal deposits found in the southeastern corner of the state. During the Permian Period, the rocks underlying the Flint Hills were deposited. Other more recent periods of geologic history were the Cretaceous and the Tertiary.

Most periods were named for the broad geographic localities where their deposits were first recognized and described. For example, the Mississippian was named for the typical outcrops in the Mississippi River valley; the Pennsylvanian, for outcrops in that state; the Permian, for the province of Perm in Russia. Other rocks have names derived from other sources. Cretaceous means chalky in Latin and was so named because chalk beds are sometimes found in rocks of that age.

Geologists refer to the rocks deposited during a specific time as belonging to that age. The chalk deposited in western Kansas during the Cretaceous, for example, is Cretaceous-age rock, or part of the Cretaceous system. Similarly, the coals of eastern Kansas are part of the Pennsylvanian system. By using standard names for the different geologic periods, all geologists know the age of the rocks referred to, no matter where they are found.

During those geologic periods in Kansas, layer after layer of sediments were deposited. In some places, such as the southwestern corner of the state, those sedimentary deposits are as much as eight-thousand feet deep. Those layers can be thought of as beds of rock that extend for great distances. The layers may vary in thickness from place to place, and they may be slightly bent or torn or otherwise disturbed by natural forces. But geologists can still identify and trace them for hundreds of miles.

To help them study and discuss those different layers of rock, geologists have developed a system for naming them. All of the rock layers have been divided into much smaller units called formations and members. The sedimentary rocks of Kansas, for example, have officially been divided into approximately 170 formations, each named for the geographic locality where it was first recognized and described. The Oread Limestone Formation, for example, was named in 1894 by Erasmus Haworth—the first director of today's Kansas Geological Survey—for outcrops on Mount Oread where the University of Kansas is located. That same Oread Formation can be seen in exposures along the Interstate highway near Lawrence.

Many formations are divided into smaller units called members, also named for geographic locations. The Oread Formation is divided into seven such members, one of which is the Leavenworth Limestone Member, a foot-thick bed that can be followed for hundreds of miles where it crops out and is visible at the surface. Although the Leavenworth Limestone Member was named for outcrops near that city, it retains the same name no matter where

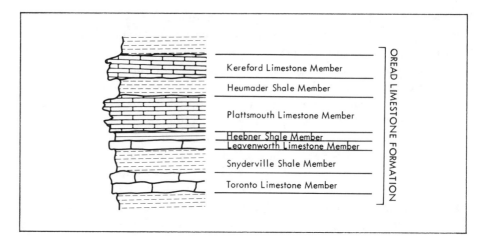

Figure 2—A sketch of the Oread Limestone Formation shows its seven named subdivisions, or members. These beds occur in the same order wherever the Oread Formation is found in Iowa, Missouri, Kansas, and Oklahoma.

else it is found in Kansas. Therefore, each layer of rock has a name and each has its own place in the succession of rock layers that make up the state's subsurface.

Every one of those 170 rock layers in Kansas has its own story to tell about the geologic history of the state. Most of that story is a description of two natural processes, erosion and deposition, which act continuously to modify the earth's surface. In Kansas, the rock layers largely tell the story of land that has been covered with seas and then lifted above the ocean's surface. As soon as any part of that land appeared above the water, the forces of erosion went to work to wear the land back down to sea level. The material that was eroded from the land was carried by streams and rivers back to the sea where it was deposited and buried, compressed into new layers of rock.

The present face of Kansas shows the effect of such erosion. Thousands of feet of rock have been weathered away at the extreme southeastern corner of the state, and westward from that point successively younger layers of rock are exposed in overlapping strips like steps up a flight of stairs. As a result of the steplike erosion, motorists driving across the state from southeast to northwest travel through successive outcrops that record 325 million years of geologic history. As a general rule, in Kansas the rocks get younger as you travel from east to west. The oldest rocks—at least those present at the surface of the state—are found in the eastern part of Kansas. The formations to the west are successively younger, so that the rocks and soils in western Kansas are considerably less aged than those in the east.

Western Kansas is also considerably higher. In traveling from southeastern Kansas to the northwestern corner of the state, travelers gain more than half a mile in altitude. South of Coffeyville, in Montgomery County, the elevation is only about seven hundred feet above sea level, but by the

Colorado border the elevation has reached more than four thousand feet. That increase would probably be dramatic if it were not spread out over the four-hundred mile length of the state; instead, the change in elevation in Kansas is gradual, and few travelers even notice it. That change in elevation is caused by additional layers of sediments that blanket western Kansas but are not present in eastern Kansas. Most of that additional sediment in the west is the result of erosion of the Rocky Mountains, material that has been transported across Kansas by water and wind. Deposition has buried the Pennsylvanian rocks in western Kansas that appear on the surface of eastern Kansas.

Viewed in this light, then, a drive across the state is not such a mundane undertaking as it might seem to the uninformed. Driving from east to west across Kansas means gaining more than half a mile in altitude. It means crossing the bottoms of ancient oceans and traveling through areas that were once covered by miles and miles of salt flats. Most importantly, driving across Kansas reveals the handiwork of nature over the course of more than 300 million years. If we take the time to read it, the story of those past geologic epochs lies in the rocks all around us.

Generalized Physiographic Map of Kansas

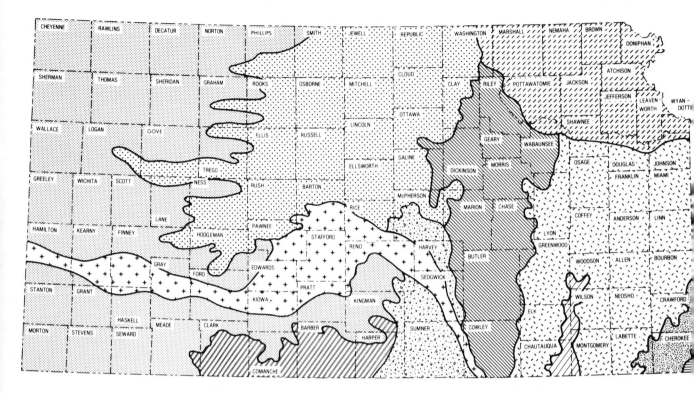

Figure 3—The map shows the topography of the eleven different regions of the state.

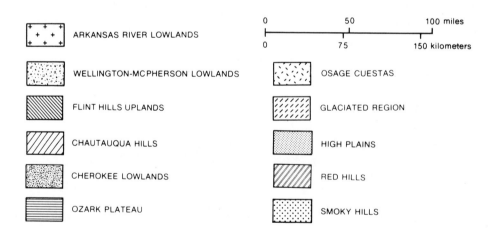

- ARKANSAS RIVER LOWLANDS
- WELLINGTON-MCPHERSON LOWLANDS
- FLINT HILLS UPLANDS
- CHAUTAUQUA HILLS
- CHEROKEE LOWLANDS
- OZARK PLATEAU
- OSAGE CUESTAS
- GLACIATED REGION
- HIGH PLAINS
- RED HILLS
- SMOKY HILLS

```
0          50          100 miles
|----------|-----------|
0          75          150 kilometers
```

1

Landscapes: A Geologic Diary

by Frank Wilson

The easiest way to learn about the geology of Kansas may be to begin with those features that are visible at the surface, such as the Flint Hills, the Red Hills, the chalk beds of western Kansas—all the geologic features that make up the Kansas landscape. In short, this chapter is a geological sampler, describing the eleven different physiographic regions that comprise the landscape (see Figure 3).

To begin, think of the rock layers in Kansas as the pages in a book, with the oldest layers on the bottom and the newer sediments on top (see Figure 4). These rock layers make up the geologic history of Kansas; every layer is like a page in the geologic diary of the state, each recording a moment in the geologic development of Kansas. By examining each layer where it is exposed at the surface, we can piece together the story behind the Kansas landscape as it appears today. This chapter, then, is a look at the geologic diary of the state, beginning with the oldest rocks and moving on to the younger parts of the state's landscape. That diary begins in southeastern Kansas.

Mississippian Rocks and the Ozarks of Kansas

The oldest strata exposed at the surface in Kansas are those of the upper part of the Mississippian system. These rocks were deposited 330–360 million years ago. They now crop out in a small triangle at the extreme southeastern corner of the state.

The Mississippian sediments are limestones containing bedded chert or flint. Because chert is much harder and more resistant to weathering than limestone, erosion of the softer limestone has left a thick blanket of chert gravel and soil on the tops of all the higher ridges and hills. The cherty limestones themselves are only exposed in road-cuts or steep cliffs along

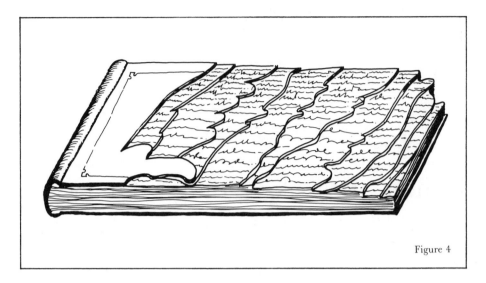

Figure 4

stream valleys. One of the most beautiful of these outcrops is a cliff on the north side of Shoal Creek about one and a half miles south of Galena on State Highway 26. The particular limestone exposed in this cliff contains few chert layers. Crinoids, trilobites, and other fossils that are typical of Mississippian-age rocks may be found at this site. If the scenery here resembles the Ozarks, it is because this fifty-five square-mile area in the southeastern corner of the state is a part of the Ozark Plateau physiographic region.

Before being buried by thousands of feet of younger sediments, these same rocks were at the surface of the earth at the close of Mississippian time (some 330 million years ago). The landscape may even have looked much the same as it does now. Swiftly flowing streams had notched themselves deeply into limestone bedrock and in some places flowed through caverns that had been carved into the soluble limestones by running water. In some places the roofs of caverns collapsed to form large sinkholes at the surface of the ground. When the seas returned during the Pennsylvanian period and again covered the region, the first muddy sediments filled the caves, sinkholes, and stream channels. Millions of years later, while Mississippian rocks were still covered by hundreds of feet of younger sediments, the brittle, cherty limestones were crushed and broken by a massive shifting of the crust of the earth. Mineralized water moved through the broken rock and deposited zinc, lead, and other ores in openings in the rock. Because the mineralized water could not readily penetrate the shale that filled the caverns or sinkholes, the lead and zinc ores were commonly deposited in richest concentrations along the walls of these structures. The first prospectors and miners in the Galena district sought out these rich ore zones in caverns and sinkholes, which they called runs and circles.

Lead and zinc mining in the Galena area reached a peak in the late 1890s. As the shallow ores were mined out, prospecting shifted to deeper ores in the area west of Baxter Springs. The huge piles of chat (cherty rock from which lead and zinc ores have been removed) dotting the countryside southwest of Baxter Springs are evidence of the size and extent of these underground workings. Literally thousands of trainloads of chat, or tailing, have been hauled away to be used as roadbed material for railroads all over the United States, but small mountains of it still remain (see Figure 6).

While lead and zinc mining has stopped, it still affects many residents of southeastern Kansas. Studies show that twenty-two hundred acres of Cherokee County have been undermined, and those old shafts occasionally collapse, endangering people and property. The old mines have also contributed to ground-water contamination in the area.

Pennsylvanian Rocks

The next chapter of the geologic diary of Kansas is written in rocks of the Pennsylvanian system. Coal is so common in deposits of Mississippian and Pennsylvanian ages all over the world that in countries other than the United States, these two systems are combined into a single system called the Carboniferous (containing coal).

Cherokee Coal-Bearing Strata and the Cherokee Lowlands

The oldest strata of Pennsylvanian age occurring at the surface in Kansas are the coal-bearing strata of the Cherokee Group. The shales, sandstones, and coals of the Cherokee Group are easily eroded, and the land surface in their outcrop belt is fairly flat and poorly drained. This plain is called the Cherokee Lowlands. After the major period of erosion that closed Mississippian time, the region that is now eastern Kansas stayed nearly at sea level for a very long time. Great swamps covered the low-lying areas around the margins of the seas, and primitive plants (including ferns as large as today's trees) grew densely in the bogs. Dead vegetation accumulated at the bottoms of the swamps and was later changed to coal by deep burial below a few thousand feet of younger sediments. The Cherokee strata of southeastern Kansas contain at least fourteen coal beds, twelve of which are thick enough to have been mined at various times in the past. The first settlers found the coal beds at the surface of the ground and "mined" them simply by breaking coal from the outcrop and loading it onto wagons. It was not until railroads were built into the area, however, that coal mining became an important industry in southeastern Kansas. Coal played a key part in the development of railroading during the period immediately before and after the Civil War. Because coal was a more efficient and less bulky fuel than wood, it was soon used almost exclusively in the steam locomotives of that day. To meet the demands of the railroads, a number of strip mines were opened during the 1870s in the shallow coals in Bourbon, Cherokee, and Crawford counties. Miners' camps sprang up near these mines, and today many of the towns of the area bear the colorful names of the old coal camps. Pittsburg, now the largest city in the area, was just another of these small coal towns until the zinc smelters were established there following the

Figure 5—This outlier of Mississippian-age rock is southwest of Galena, Cherokee County.

Figure 6—Spoils piles from lead and zinc mining in Galena, Cherokee County.

lead and zinc boom at Galena in 1878. The first shaft mine in southeastern Kansas was dug in Cherokee County by the Scammon brothers in 1874. Despite warnings from their friends that underground mining would not work at such shallow depths, the mine soon developed a capacity of forty carloads of coal a day. Within a few years, underground mining became the principal method of coal production in southeastern Kansas coal fields.

Underground mining continued in southeast Kansas until the late 1950s, when it gave way again to strip mining. Beds of coal that are too thin to be mined underground can sometimes be profitably stripped by power shovels, some of which are now able to dig to depths of more than a hundred feet. These monster shovels dig a wide trench down to the top of the coal, which is then dug up and loaded into trucks by smaller shovels (see Figure 7). At the end of each trench, the big shovels turn around and move back in the opposite direction, piling the stripped soil into the trench that was just finished. When the shovels finally move on, the lands looks like a giant

Figure 7—No longer used, this huge shovel once strip-mined coal south of West Mineral, Cherokee County.

Figure 8—Southeastern Kansas coal strip-mines, filled with water, are now used for fishing; located in Cherokee County.

plowed field in which each of the ridges and furrows is perhaps a hundred feet wide and fifty feet deep.

The chief use of coal today is as fuel for steam-powered generation of electricity. New generating plants are being built near areas of coal reserves because it is more economical and efficient to transmit electricity by power lines than to haul coal to generating plants at central points. As the public's use of electrical power increases, more and more of the earth will be dug up to supply the coal to meet the demand. Only a few years ago this stripped ground was considered ruined, but a new concern for the earth and its proper management for future generations has changed this view and efforts are now being made to reclaim the land. Since 1969, Kansas legislation has required coal companies to level the newly stripped land to a rolling terrain and plant it with trees or grass.

Many years before the general public was aroused to the need to preserve the environment, a few coal companies and several private landowners in southeast Kansas had proved that the stripped ground could be leveled, planted with grass, and returned to productive use as grazing land. Following their lead, and with financial aid and technical assistance from state and federal agencies, other private landowners have made a start toward reclaiming the forty thousand acres of strip-mined lands in Cherokee and Crawford counties.

Until generation of electrical power by nuclear energy is developed as a safe and practical process or other alternatives become economical, coal will continue to play an important part in the economy of Kansas, and efforts to convert the formerly wasted, stripped lands into a useful resource will continue.

Coal in Other Parts of Kansas

The outcrop of the Cherokee coals extends northeastward through Missouri into eastern Iowa and western Illinois. Coal had been mined in those areas for a number of years before it was extensively developed in southeastern Kansas. Major Frederick Hawn, an early military surveyor and geologist in Kansas, had studied these coals and the geology of the area between their outcrop and eastern Kansas. On the basis of his studies, Major Hawn predicted that the Cherokee coals existed at a depth of about seven hundred feet below the city of Leavenworth, Kansas. As early as 1859, Hawn organized an exploration company and with a crude drill began to sink a hole to prove his theory. His first efforts were interrupted by the Civil War and financial difficulties, but in 1865 a two-foot thick seam of coal was found in the Cherokee shale at a depth of 713 feet. The right to mine under the Fort Leavenworth military reservation was granted by the U.S. Congress in 1868, and by 1870 a shaft had been sunk to the coal. A hole drilled from the bottom of this shaft showed four other minable coals at greater depths. During the 1880s and 1890s at least four other mines were opened in these seams near Leavenworth, including one operated by the state of Kansas at the penitentiary in Lansing. By the end of the 1890s about thirteen hundred men were employed in the various deep mines of Leavenworth County.

Although the most productive coals in Kansas are found in the Cherokee Group of the Pennsylvanian system, other coals are scattered through the remainder of the Pennsylvanian and lower Permian systems. There are few counties in eastern Kansas that have not reported coal mining at some time in their histories.

Osage Cuestas of Eastern Kansas

The next pages in the geologic diary of Kansas are represented by the alternating limestones and shales of the upper Pennsylvanian and lower Permian systems. These strata form the bedrock of the Osage Cuestas physiographic division. In this area, successively younger beds overlap one another toward the west, like tilted steps of a giant staircase. The limestone formations are more resistant to weathering than are the shales occurring between them and, because of that, the limestones crop out as the tops of the steps while the thick shales form the risers between steps (see Figure 9). All beds slope gently toward the west-northwest, so the rise in the ground surface from one step to the next is not great. The dipping strata form a series of parallel ridges having gently sloping west faces and steeply sloping east faces. This type of landform is called a cuesta.

The limestones and shales visible in the Osage Cuestas region of Kansas were deposited in shallow seas that lapped onto the central part of our continent some 250–300 million years ago. The type of rock exposed in a particular outcrop gives some indication of the nearness of the shoreline at the time that particular layer was deposited. As material was eroded from the land and carried into the sea by streams, the coarsest material was deposited near the shore while finer sediments were carried farther seaward. Although there are exceptions to this general rule, a bed of sandstone usually indicates deposition on or very near shore; layers of shale (which were once mud) indicate deposition a little farther from shore; and limestone

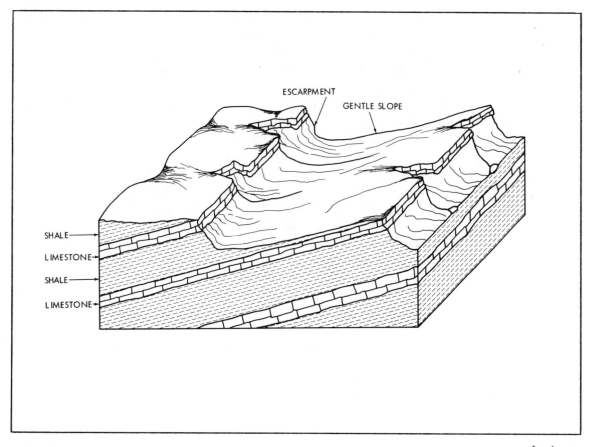

Figure 9—Cuesta topography developed in gently dipping, alternating hard and soft strata.

usually indicates deposition in the open sea, far from land. Years ago geologists who studied the Pennsylvanian and Permian strata of the midcontinent noted that sandstone, shale, and limestone reoccurred in predictable order upward through a vertical sequence of beds. In its simplest form, this recurring order was: sandstone, shale, limestone, shale, and sandstone. Geologists theorized that such a sequence represented a single advance and withdrawal of the shoreline of an ancient sea past a certain geographic point. Dozens of such cycles were soon recognized in the strata of the Osage Cuestas region. From this and other evidence, geologists deduced that the Pennsylvanian and Permian seas were relatively shallow and that their bottoms were nearly flat or only gently sloping. Because of this, small changes in sea level caused the shoreline to move great distances back and forth across the region that is now Kansas. The knowledge that each ledge of limestone is a cross section of the bottom of an ancient sea—and that the shales and sandstone exposed in road-cuts are evidence of ancient environments now alien to Kansas—may make a trip across our state more interesting.

Along with abundant resources of natural gas at shallow depths, the limestones and shales of the Osage Cuestas region were once the basis of a thriving mineral industry in southeastern Kansas.

Limestone is the primary ingredient in the manufacture of Portland cement. The rock is finely ground, mixed with shale and other natural materials, then baked at high temperatures in rotary kilns to produce cement. In a somewhat similar manner, shales are ground, mixed with water, molded, then baked in ovens to produce brick or tile. During the late 1890s and early 1900s there were fifteen cement plants and sixty brick or tile plants operating in southeastern Kansas.

Because of the seemingly limitless supply of natural gas, other industries were also attracted to the area. Ten zinc smelters were opened in the Iola area alone, and eight others were scattered through the gas fields of Allen, Neosho, Wilson, and Montgomery counties. At about this same time, the depletion of natural-gas reserves in Indiana caused a wholesale migration of glass plants from Indiana to Kansas. Between 1890 and 1930, twenty-five glass plants were moved to or established in southeastern Kansas. In many instances entire companies, their equipment, workers, and families were moved en masse to their new locations. The supply of natural gas soon waned, however, and within a few decades the industries that depended on the gas began to fade away. Most of the glass plants and their highly skilled artisans moved on to new areas of abundant gas supply in Oklahoma or Texas. Some of the brick plants and cement plants were converted to coal or oil and continue in operation to the present day, but many others were shut down and abandoned.

Some of the buildings that once housed brick, tile, or glass plants are now used for other purposes, but the massive concrete foundations of abandoned cement plants stand silently in the countryside, grown over by trees and weeds.

Chautauqua Hills

Extending northward into the Osage Plains from the southern boundary of Kansas is a sandstone-capped, rolling upland called the Chautauqua Hills (see Plates 4 and 5). The thick sandstones underlying this area are the remains of ancient deltas deposited by streams as they emptied into the Pennsylvanian seas. Dense growth of jack oak and post oak blanket the sandstone hills, and blossoming redbuds abound along the wooded valleys during the early spring. Travelers can enjoy this beautiful scenery by following the Redbud Trail through the hills near Sedan.

Permian System

The next chapter in the geologic diary of Kansas is written in Permian rocks that were deposited some 240–290 million years ago.

Flint Hills

The Flint Hills stretch across the width of Kansas along the western edge of the Osage Plains. Although the Flint Hills were formed by erosion of gently westward-dipping strata in much the same fashion as the Osage Plains, the Permian-age limestones of the Flint Hills contain numerous bands of chert, or flint. Because chert is much less soluble than the limestone that enclosed it, weathering of the softer rock leaves behind a clayey soil containing much flinty gravel. This gravelly soil blankets the rocky uplands and slows the process of erosion compared with the rate of weathering in adjacent areas where the limestone bedrock does not contain chert. As a result, the crests of the Flint Hills are at a higher elevation than the territory directly to the west or east and, as topographic land forms, the hills themselves are older than those of adjacent areas.

Some Kansans call these grassy uplands the Blue-Stem Flint Hills, and it is certainly true that the combination of native blue-stem prairie grass, topography, and geology makes the region unique. The virtual absence of trees on the uplands of the Flint Hills has been attributed both to the effects of periodic droughts and to lightning-set prairie fires that undoubtedly swept the plains for thousands of years before even the ancestors of the American Indian ventured onto this continent (Plates 6 and 7). In more recent times, both the Indians and the white settlers purposely set fire to the prairie to burn off the dry grass. Because the prairie grasses are dormant during seasons when fires are most likely to occur, and can quickly regenerate from roots or seed when normal weather returns after a drought, they are well adapted to conditions on the plains. Trees, on the other hand, are restricted to the bottoms of valleys and to sharp breaks in topography where they are sheltered from fire and are able to find sufficient ground water to survive prolonged drought. Thus, with the aid of fire and drought, the grasslands remained stable for nearly ten thousand years. When settlers came to the plains in large numbers during the 1870s, the prairie extended from eastern Kansas westward to the Front Range of the Rocky Mountains, northward to the Dakotas, and southward into Texas. Within a few decades after the arrival of settlers, much of this vast grassland was plowed up. The tall-grass region of the Flint Hills was at the eastern margin of the prairie and would have been the first area to be plowed except that the flinty gravels soon convinced the sodbusters that the land was not suitable for cultivation (Plate 8). Many of them moved on to the west in search of better land, but a few resourceful pioneers recognized that it was easier to work with Nature than against her. They built their farmsteads in the sheltered valleys where water was present even in times of drought and turned their cattle to graze on the grassy uplands. The Flint Hills region still remains largely untouched by the plow and is the last large preserve of native tall-grass prairie in the United States.

Most major east-west highways in Kansas pass through the Flint Hills, and the Kansas Turnpike cuts diagonally through the heart of the hills between El Dorado and Emporia. It is difficult, however, to feel the mood of the hills while traveling through them at top speed in an automobile. To recapture the awe that must have gripped the first explorers as they viewed this vast sea of grass, the traveler must leave the busy routes and wander deep into the hills along the gravel roads that wind through the region (Plates 10 and 11). One such marked route is the Skyline Scenic Drive south of I-70 between the Snokomo interchange west of Topeka and the Wamego–Mill Creek interchange east of Manhattan. Another beautiful but unmarked route is from the Cassoday exit of the Kansas Turnpike eastward to Teterville and then generally northeastward by gravel road toward Emporia. The Flint Hills can be traversed from north to south for almost their entire length by following State Highways 99 and 13 from the north boundary of Kansas southward to Manhattan and State Highway 177 from Manhattan south to El Dorado.

Wellington-McPherson Lowlands

The type of Permian sediments deposited after the cherty limestones of the Flint Hills indicate that the ancient climate became very dry for a long period of geologic time. As the Permian seas shrank in size, shallow arms of these seas were cut off from the open ocean, causing great thicknesses of dark shale, gypsum, and salt to be deposited as the waters evaporated. The Wellington-McPherson Lowlands of central Kansas are underlain by such sediments. The beds of salt and gypsum have been dissolved by fresh ground water near the outcrop and are seldom exposed at the surface, but great thicknesses of these sediments are present below much of western Kansas. A three-hundred-foot–thick salt bed is mined at depths of seven

Figure 10—The Red Hills in Barber County, also called the Gypsum Hills.

hundred feet near Hutchinson and at a depth of a thousand feet near Lyons. Salt beds below the surface in far-western Kansas occur in three horizons, totaling more than seven hundred feet in thickness.

Valuable national records and documents are stored in deep salt mines at Hutchinson. Because salt is extremely stable where it is deeply buried, the U.S. Atomic Energy Commission at one time suggested that the salt beds of central and western Kansas be used for storage of radioactive wastes.

Red Hills

During the latter part of the Permian period several thousand feet of brick-red shales, siltstones, and sandstones were deposited in the area that is now Kansas.

Along the southern border of Kansas, the Permian red beds have been exposed by erosion in the Red Hills physiographic region. Beds of light gray gypsum and dolomite cap the shale hills to make a ruggedly beautiful, butte-

and-mesa topography unlike any other in Kansas (Plates 15 and 16). The Plains Indians called these the Medicine Hills and the major stream that flows from them Medicine River. They believed that spirits in the hills and streams helped to cure their illnesses and hastened the healing of wounds. In a sense they were right, because the water of the springs and streams contains calcium and magnesium sulfates and other natural salts dissolved from the gypsum and dolomite beds that abound in the headwaters. Many of these natural chemical compounds have therapeutic and healing effects. Magnesium sulfate, for example, is known to pharmacists as Epsom salt. Before the discovery of antibiotics, solutions of Epsom salt were often used to draw infection from wounds and to promote healing. In very dilute form, it was used as a tonic in the same manner as sulfur and molasses. So what Indians long ago discovered and attributed to friendly spirits, geologists and chemists later confirmed by observation and analysis.

Sinkholes

In Meade and Clark counties, just beyond the western boundary of the Red Hills, dissolution of the massive gypsum and salt beds several hundred feet below the surface is believed to have caused a number of large sinkholes. Big Basin and Little Basin, two well-known sinks in western Clark County, apparently formed in the recent geologic past but before historical record (Plates 17 and 18). Big Basin is about a mile in diameter and more than a hundred feet deep. U.S. Highways 160 and 283 pass directly through it about two miles south of their junctions in northwestern Clark County. Little Basin is located about a third of a mile east of the east rim of Big Basin. A pool of water in Little Basin is known as St. Jacob's Well (see Figure 12).

The Meade Salt Sink formed suddenly in March 1879, about one and a half miles south of the city of Meade, and swallowed a part of the Jonas and Plummer Trail, a much-used wagon and cattle trail of the period. The hole was about sixty feet deep and 610 feet in diameter. It gradually filled with salt water to within about fifteen feet of the ground surface. At one time, salt was produced commercially by evaporating the water from this sink. It is now partly filled with sediment and usually is dry.

In April 1973, a new sink began forming about one-half mile northeast of the old Meade Salt Sink. This sink has been named the Seybert Sink.

Englewood-Ashland Lowland

In addition to these well-known sinkholes, geologists believe that the extensive lowland north of the Cimarron River in Clark and eastern Meade counties was formed by dissolution of the deep gypsum and salt horizons, which resulted in sinking of the ground surface over wide areas. These sinkholes were incorporated in the Cimarron River drainage system by normal surface erosion, which eventually smoothed the floor of the depression into a plain several hundred square miles in extent. According to this

Figure 11—A butte in the Red Hills west of Medicine Lodge, Barber County.

theory, both Big Basin and Little Basin are a part of the continuing solution-subsidence process and will eventually be swallowed by enlargement of the Englewood-Ashland Lowland.

Triassic and Jurassic Systems

Rocks of Triassic age are not recognized in Kansas. Strata thought to belong to the Jurassic system are found in the subsurface and at the surface in only two small outcrops along the Cimarron River north of Elkhart in Morton County at the extreme southwestern corner of the state (see Figure 13).

Cretaceous Rocks and the Smoky Hills

The next major chapter in the geologic diary of Kansas is written in rocks of the Cretaceous system. Chalky sediments are typical of this system of rocks throughout the world. The famous White Cliffs of Dover on the eastern coast of England and the chalks of the Paris basin, for example, are of the same general geologic age as the chalks of western Kansas.

The outcrop belt of these rocks is the Smoky Hills physiographic region. The front of these uplands can be seen for a considerable distance as they are approached from the east. They were probably called the Smoky Hills by early settlers because they are often partly obscured by atmospheric refraction and haze. The Smoky Hills are actually made up of three levels of uplands, each capped by a successively higher and younger group of erosion-resistant rocks.

Dakota Hills Country

The easternmost range of hills is capped by thick sandstones of the Dakota Formation. The eastern edge of this region is marked by outlying hills and buttes that rise sharply above the surrounding plains. These high

Figure 12—St. Jacob's Well in a natural sinkhole called the Little Basin, located in western Clark County. Nearby is a larger subsidence feature called the Big Basin.

Figure 13—The view from Point of Rocks in Morton County, extreme southwestern Kansas. The rocks, part of the Ogallala Formation, were an important landmark along the Santa Fe Trail. Beneath the Ogallala, rocks of Jurassic age crop out; this is the only part of Kansas where Jurassic rocks occur at the surface.

Figure 14—An outcrop of Dakota Sandstone, in the Smoky Hills of northeastern Ellsworth County.

hills furnished excellent vantage points both for early explorers and for raiding parties of warlike Indians who sought to protect their hunting grounds from encroaching settlers. Pawnee Rock in Barton County and Coronado Heights north of Lindsborg in Saline County are two such historic sites. Pawnee Rock was a famous landmark on the Santa Fe Trail. It was once considerably higher than it is now, but much of the sandstone was quarried in earlier days for building stone. It was named for a band of Indians besieged there by a larger war party of another tribe.

Coronado Heights is so named because it presumably was on the route of Coronado, the Spanish conquistador who explored this region in 1541 searching for Quivira, the fabled City of Gold. Coronado is thought to have climbed to this vantage point to survey the surrounding plains.

Rock City, south of Minneapolis in Ottawa County, and Mushroom Rocks State Park, south of Carneiro in Ellsworth County, are areas of unique geologic interest in the Dakota Sandstone country. At Rock City, huge spheres of sandstone, some as large as twenty-seven feet in diameter, dot the surface like giant bowling balls (see Figure 15). At Mushroom Rocks State Park, similar rocks are balanced on natural pedestals of softer rock (Plate 14). The spheres are caused by differential erosion of the Dakota Sandstone. Both of these areas were once covered by a considerable thickness of sandstone. Ground water circulating through the sandy rock deposited a limy cement that grew outward in all directions from calcite crystals or limy fossil fragments scattered throughout the sandstone. As the softer uncemented portions of the sandy rock were weathered away by erosion, these spherical zones of cemented sandstone remained. Although some of the giant concretions have toppled from their pedestals, bedding exposed on the sides of the spheres indicates that most of them remain in their original positions.

Small fist-size sandstone concretions from the Dakota Formation are commonly found around the shoreline of Kanopolis Lake. Many of these concretions are hollow inside and look like crudely shaped pots. Others, when broken, are found to have a shiny metallic interior made up of pyrite, or fool's gold.

The beautifully rugged topography that characterizes the Dakota Sandstone country is nowhere better displayed than in the vicinity of Kanopolis State Park and Lake. This area rivals the Flint Hills for the beauty of its natural prairie.

Much of the Dakota Formation was deposited on land, and fossils of land plants and ancient animals are often found in the sandstone. Beds of lignite, or brown coal, in the upper part of the formation are evidence of deposition in shoreline swamps, which marked the return of marine waters that covered central and western Kansas in later Cretaceous times. The Graneros Shale, which overlies the Dakota Formation, was deposited in gradually deepening seas as the shoreline moved northward and westward across the western part of the state. Where the dark gray Graneros Shale is

Figure 15—Large concretions, composed of Dakota Sandstone, at Rock City in Ottawa County, southwest of Minneapolis. *to 27' diameter*

not thickly covered by vegetation, the exposed slopes are often littered with large prisms of selenite, or crystalline gypsum.

Fencepost Limestone Country

The middle range of uplands in the Smoky Hills physiographic region is capped by interbedded thin chalks and chalky shales of the Greenhorn Limestone. The base of the Greenhorn Limestone is marked by a thin, sugary-appearing limestone bed. By carefully searching outcrops of this limestone, amateur collectors may find well-preserved sharks' teeth and remains of other marine fossils. Considering the ferocity of present-day sharks, the abundance of sharks' teeth in this thin deposit causes one to wonder how any other creature survived in the Cretaceous seas.

The Greenhorn Formation is made up of thin, chalky limestone beds (usually less than a half-foot thick) alternating with thicker beds of grayish shale. Fossil clams and oysters are quite common in the thin limestone beds. One particular one-foot thick limestone near the middle of the Greenhorn is so full of well-preserved clams and oysters that it is known as Shell Rock.

Figure 16—A stone fencepost sits atop the Fencepost Limestone bed, near Wilson Reservoir in Russell County.

At the very top of the Greenhorn Formation is another one-foot thick layer of rock called Fencepost Limestone (Plates 12 and 13). When settlers first came to the Smoky Hills region, native timber for fenceposts and lumber was hard to come by on the nearly treeless plains. Because of this, both the Shell Rock and the Fencepost Limestone strata were extensively quarried for masonry stone and for rock fenceposts (see Figure 16). The Fencepost bed was favored for quarrying because it was more uniform in thickness than the Shell Rock. The soil was removed from the ledge and a line of holes was drilled into it about a foot back from the edge. Wedges were then fitted between feathers (rounded, tapered metal strips) into the holes and were gradually and uniformly tightened by tapping each with a sledge hammer and listening to its ping. By carefully tapping the wedges, a slab of limestone could be split off and sawed into the proper length for a fencepost. The chalky rock was relatively soft when first quarried and could be drilled and sawed by hand, but after being set in place and curing, the rock fenceposts became much harder. Many rock fenceposts in this region were first set in place nearly a hundred years ago. Although several generations of barbed wire have rusted away during that span of time, the posts remain in good condition.

Figure 17—Rock fenceposts, Russell County.

Figure 18—This bridge, in the Hell Creek area south of Wilson Reservoir, was built of Fencepost Limestone and Dakota Sandstone.

Figure 19—Monument Rocks, composed of Cretaceous-age chalk, stand thirty to forty feet above the Smoky Hill River valley in Gove County.

Figure 20—Different rock strata are evident in Monument Rocks, Gove County.

Figure 21—Erosion of chalk formations in southeastern Logan County produces a badlands appearance.

The third and westernmost range of hills in the Smoky Hills physiographic province is developed on the thick chalks of the Niobrara Formation of late Cretaceous geologic age (see Plates 1, 19, 20, and 21).

Chalk is a sediment of the open ocean, composed almost entirely of the remains of microscopic marine plants and animals that floated or swam near the surface of the Cretaceous sea. So numerous were these tiny organisms that their remains continuously settled to the sea floor where they formed a soft limy ooze that swallowed up and preserved the remains of larger animals that sank to the bottom. Preserved in the chalks of western Kansas are fossilized skeletons of extinct fishes and giant flying or swimming reptiles. Excellent examples of these fierce-looking toothed vertebrates are displayed in the Sternberg Memorial Museum at Fort Hays State University in Hays.

The thick chalk beds of the Niobrara Formation are best exposed in bluffs of the Solomon, Saline, and Smoky Hill rivers west of a line drawn diagonally northeastward from Finney County to Jewell County. Two unique and often-photographed spots in the Niobrara outcrop area are Castle Rock in eastern Gove County and Monument Rocks in western Gove County (Figures 19 and 20). At both of these places, the ancestral Smoky Hill River carved its course through the outcrop of the Niobrara chalk and left behind these naturally sculptured erosional outliers. They now stand alone on the prairie like abandoned remnants of some ancient civilization. Both of these sites are located on private land, and visitors are welcome, as long as they are careful not to deface the monuments or to litter the grounds.

The Tertiary System and the High Plains

The great mountain systems of western North America were born near the end of Cretaceous times. During the Tertiary period that followed, the mountains continued to be slowly uplifted by massive movements of the earth's crust. Streams flowing eastward across eastern Colorado and western Kansas were choked with rock debris eroded from the newly formed Rocky Mountains. So great was the mass of eroded material that it literally overflowed the stream valleys and spread out over the uplands. By the end of the Tertiary period (which lasted for about 60 million years), the upper surface of this immense sheet of sand and gravel formed a gently eastward-sloping plain extending from the eastern front of the Rockies to the west slope of the Flint Hills in central Kansas. The High Plains of western Kansas are the uneroded remnants of this extensive plain, and the great wedge of sand and gravel that lies below the surface is the Ogallala Formation.

The Ogallala is covered by a veneer of younger sediments on the uplands, but it is often exposed in the bluffs of the streams that cut through it (Plate 23). In many places, the sand and gravel of the Ogallala contain caliche, a soft limy cement deposited by evaporating ground water. These zones are called mortar beds because the limy material could be calcined by heating it over open fires; it was used by early settlers as mortar when building their houses of native limestone. Outcrops of Ogallala mortar bed are beautifully exposed in the bluffs around Scott County State Lake.

Arkansas and Cimarron River Deposits

The Arkansas and Cimarron rivers are present-day analogs of the Tertiary-age streams that deposited the Ogallala Formation. Both the Arkansas and Cimarron originate in the Rocky Mountains, and their valleys are filled with sand and gravel eroded from those mountains during Pleistocene or more recent times. Except in times of flood, the water is

Figure 22—A scene in the High Plains region, near Grainfield in northern Gove County.

Figure 23—The Ogallala Formation, near Scott County State Lake in northern Scott County.

barely visible above the surface of the valley-filling sediments, and much of the flow is through the sandy deposits themselves. The great expanses of sand that form the Arkansas River Lowlands and the Great Bend Lowlands and other deposits south of the Cimarron River in southwest Kansas are similar to those that make up the Ogallala Formation. If you can imagine the Arkansas and Cimarron rivers meandering back and forth across Kansas for another 60 million years, leaving in their abandoned channels the reworked sediments of the eroding Rockies, then you can visualize the processes that formed the High Plains.

The huge supplies of ground water contained in Tertiary and Pleistocene deposits of western and southwestern Kansas are among the most important natural resources in the state. Like oil and gas, these supplies were once considered to be limitless, but hydrologists now know that water is being pumped out for irrigation and other uses faster than it is being naturally replenished.

Figure 24—Large quartzite boulder, carried into northeastern Kansas by glaciers, located north of Vermillion in Marshall County.

Quaternary System

Pleistocene Epoch

The last and perhaps unfinished chapter of the geologic diary of Kansas was written during the Pleistocene epoch.

The Pleistocene is often called the Great Ice Age because at its climax nearly a third of the land surface of the earth was covered by glaciers, or thick sheets of slowly moving ice. Scientists do not agree as to the ultimate causes of the Ice Age, but they generally agree that winters became longer and more severe and summers became shorter and cooler. Snow that fell in the polar regions and at higher elevations of mountain ranges during the long winter did not have time to completely melt during the short summers, so it accumulated from year to year. As new layers of snow were added, the older snow was compacted into solid ice. The compressive strength of ice has certain limits; when the depth of snow and ice reached perhaps two hundred feet, the ice at the bottom of the snowpack could no longer carry the weight, and it was slowly squeezed outward or downslope from the centers of heavy snow accumulation. The year-to-year excess of snow continued for tens of thousands of years and provided the driving force to push the edges of the ice sheets farther and farther into warmer latitudes. Like slow-moving lava flows, the great glaciers that resulted overrode everything in their paths.

At least four major advances of the continental ice sheets are recognized, each separated by a period of warmer climate when the front of the ice melted back toward its source. The first glacial advance barely extended into northeastern Kansas, but the second covered much of the northeastern corner of the state north of the Kansas River and east of the Blue River. The third and fourth surges of ice did not reach Kansas, but wind-blown silt and stream-borne sediment from the melting glaciers were spread over much of the state.

Many Kansans find it difficult to imagine a part of their state covered by several thousand feet of ice, but evidence of the glaciers is abundant if travelers know where to look and what to look for. Farmers in northeastern Kansas, for example, are familiar with boulders of hard red rock scattered through their fields and pastures. The bolders are composed of red quartzite of Precambrain age, a type of rock that is not found at the surface anywhere in Kansas. Its closest outcrop is approximately two hundred miles north in southeastern Minnesota and eastern South Dakota. There, great boulders and millions of tons of other rocky debris were picked up by the ice sheets as they ground slowly southward. This debris was dumped in Kansas when the glaciers finally melted back toward the north. Such deposits are aptly called glacial till, because they were literally plowed up by the slowly moving ice. Areas in Kansas that were once covered by ice are determined by the limits of glacial-till deposits and by deep scratches in limestone and sandstone bedrock that were gouged by rocks carried along at the bottoms of the glaciers. Some of this rocky debris was carried beyond the limits of the glacial front by melting water. This material, called glacial drift, covers the higher ground in and adjacent to the area of glacial cover (Plate 9).

The valleys of major streams that flowed from the fronts of the melting glaciers were choked with sediment for hundreds of miles downstream. During the periods of warmer climate when the ice sheets withdrew temporarily to the north, the streams eroded their courses down through the level of the flood plain deposited during the advance and retreat of the preceding glacier. Because of this, stream sediments from each of the later glaciers are at lower levels than the earlier ones. The uneroded remnants of these old flood plains now form a series of broad benches or terraces along major stream valleys. Farmers often refer to the first two levels of land above the present flood plain as the second and third bottoms.

Vast areas of dry mud and barren ground were exposed as each of the ice sheets melted. Silt was picked up from these areas by the wind and spread across much of the midcontinent by dust storms that would have made the terrible storms of the 1930s seem mild by comparison. Erosion since the Pleistocene has carried away much of this material, but several feet of wind-borne silt (loess) is commonly present below the modern soil on upland areas throughout the state. Loess deposits (see Plate 22) are thickest in northern Kansas because it was close to the front of the receding glacier. More than a hundred feet of loess is exposed in the west bluffs of the Missouri River in Doniphan County. This area is sometimes called Little Switzerland because of the steeply rolling hills developed in these deposits.

The fierce Pleistocene winds swept the sands of the Arkansas and Cimarron valleys into dunes that cover thousands of square miles. Most of the dunes are now covered by vegetation but their typical hummocky shape is still apparent to travelers. An area of active dunes is present east and west of Syracuse in Hamilton County (see Figure 25).

From time to time during the Pleistocene, volcanic eruptions in the mountains to the west of Kansas spread layers of gritty, white volcanic ash

Figure 25—These sand dunes cover the landscape south of the Arkansas River, Kearny County.

over all of the central plains. The airborne ash settled over large areas during a few days or weeks and outlined features of the landscape as they existed at that particular brief moment in geologic time. Because the ash falls are like markers slipped between the pages of our ancient diary, geologists use them to correlate and date other sediments that are interbedded with them.

Many geologic time scales designate the withdrawal of the last great ice sheet as the end of the Pleistocene Epoch. However, some geologists believe that the Ice Age is not yet finished and that the earth is currently in one of the long periods of warmer climate between glacial surges. It is impossible to know whether this is true, but it is certain that the geologic diary is not yet finished and that another page is even now being written.

Because man, as an observer and recorder, has been present for such a brief moment of geologic time, it is difficult for him to perceive the exceedingly slow processes constantly at work on the earth. One of the principal concepts of geology is that the processes that affected the earth in past geologic times are the same as those at work today. For example, careful measurements show that the mountains of the western United States are rising a few centimeters per century. In this same manner, these great mountains were raised to their lofty heights. A Kansas farmer who watches his fields being gullied by spring floods is witnessing the same process that reduced the land to sea level many times before in geologic history. The key element is time, long periods of geologic time. Seemingly insignificant forces acting for eons have sculpted the landscape of Kansas.

2

Rocks

by Laura Tolsted and Ada Swineford

In 1541 the Spanish explorer Coronado made the first European observation about Kansas geology: There was no gold. Since that time rocks and minerals have played an integral part in the state's history. The earliest residents of Kansas, American Indians, used native flint to fashion their arrowheads and spearpoints; they used chunks of native sandstone to grind their grain; they even mined native clay to make pottery. The first white residents, moving into Kansas over three hundred years after Coronado, quarried limestones to build their houses and dug coal for heating and cooking. By the turn of the century, Kansans were mining everything from salt to gypsum, lead and zinc, oil and natural gas, and even using ground water for irrigation.

In short, Kansas residents have long been interested in their state's geology. That is partly due to geology's economic importance; rocks are the source for the multibillion-dollar mineral industry in Kansas. Other Kansans are interested in rocks and minerals because of their intrinsic beauty, their appearance. This chapter describes Kansas rocks, telling us where they are found, how they were formed, what they are composed of, and how we use them today. To begin, we must answer a simple question: What is a rock?

A rock is any naturally occurring mass that forms a part of the earth's crust. Such a mass may consist of sediments and particles (sand, gravel, clay, or volcanic ash, for example) as well as solid material (limestone, sandstone, granite, etc.). A rock generally is composed of one or more minerals. There are three main types of rocks, each of which was formed in a different way. These types are igneous, sedimentary, and metamorphic rocks.

Igneous rocks have been formed by the cooling and hardening of molten rock material. Igneous rocks therefore are those that have been heated to melting temperatures, forming magma, and then cooled and hardened or solidified, as black cindery rock is formed by the cooling of lava. Some rocks have been formed by slow cooling of molten materials beneath the surface of the earth. This slow cooling allows time for relatively large mineral crystals to form, and the rocks are described as coarsely crystalline. Rocks that have cooled rapidly on or near the surface of the earth do not have large crystals and are said to be finely crystalline, or glassy.

Sedimentary rocks are formed by the wearing down and dissolving of other rocks into small bits or particles of various sizes and by the later deposition, or laying down, of these particles. Such deposits, made of the broken-down substance of preexisting rocks, are called clastic rocks. There are two other important varieties of sedimentary rocks—chemical sediments (which are nonclastic) and the deposits of organic origin. The chemical deposits include gypsum, salt beds, some limestone, some quartz-containing rocks such as cherts, some iron ores, and some carbonate spring deposits such as travertine. The organically formed deposits include many limestones, diatomaceous earth, many iron ores, and coal.

These deposits may be laid down on the sea floor, along stream and river valleys, in lakes or ponds, on the land surface by the action of wind or gravity, and around the edges of and under glaciers. When first deposited, these sediments commonly are loose or unconsolidated, but many of them are hardened into solid rocks by the weight of materials above them or by the cementing action of mineral substances deposited between the grains.

Clastic sedimentary rocks are named according to the size of the particles they contain. Loose sediments in which the grains are very fine may be silt, mud, or clay. Particles slightly larger than clay size are called silt; those larger than silt are called sand; and so it goes from pebbles, to cobbles, to boulders. Of course, the sizes may be mixed, and clays and sands may be found in the same deposit with boulders, but the rock is named for the particle size that is predominant.

Metamorphic rocks are those formed by change in igneous or sedimentary rocks. This change, or metamorphism, is the result of great heat, pressure, or action of some usually hot mineral solution. Through metamorphism limestones become marble, shales become slate, and sandstones become quartzites.

Of the three main types of rocks, sedimentary rocks are by far the most common in Kansas. In fact, nearly all the surface exposures of the state consist of this type. The exceptions are several small areas of igneous rocks in Riley County and two areas of igneous and metamorphic rocks in Woodson County. Many pebbles and boulders of igneous or metamorphic rocks are found in Kansas, but they are not native to the state. They were formed elsewhere and brought to Kansas by ice or water; now they are considered as parts of sedimentary deposits.

All rocks on or near the surface of the earth are undergoing weathering. This weathering causes changes in the rock and a breaking down into smaller fragments. It may be caused by changes in temperature, by chemical action between the rock minerals and the air or ground water or plants, or by other factors. The effects of weathering can be seen on the surfaces of rocks that have been exposed to the atmosphere. The weathered surfaces commonly have a color different from that of unweathered or fresh surfaces, and one must be very careful to find the true color of the unaltered rock beneath the weathered outer surface.

Sedimentary Rocks

Limestone and Dolomite

Limestone and dolomite are two very closely related rocks. The former, in a pure state, consists of grains and crystals of the mineral calcite, and the latter the mineral dolomite (which is calcite with some magnesium added). Thus the term dolomite is used for both the mineral and the rock.

Calcite consists of calcium carbonate and is one of the most common minerals known. The mineral dolomite is made of calcium magnesium carbonate and is also fairly common. These two minerals are commonly found together in the same deposits, and whether the rock is classed as a limestone, a dolomitic limestone, or a dolomite depends on the proportion of each mineral. The two minerals can be told apart by the way dilute hydrochloric (muriatic) acid reacts on each. Cold, dilute hydrochloric acid will bubble and fizz when it is put on limestone; in order to react in the same manner with dolomite, the hydrochloric acid must be heated or the dolomite must be powdered.

In surface exposures in Kansas, dolomite is found in southeastern Kansas and in certain rocks of Permian age, which occur in the central part of the state. Limestone, on the other hand, is very widespread. Many of the rocks of eastern Kansas consist of nearly pure limestones. In the central and western parts of the state, limestone is not quite so common, although extensive deposits of chalk are found.

Limestone

Pure limestones are white or almost white. Because of impurities such as clay, sand, organic remains, iron oxide, and other materials, many limestones exhibit different colors, especially on weathered surfaces. Limestone may be crystalline, clastic, granular, or dense, depending on the method of formation; and crystals of calcite, quartz, or dolomite may line small cavities, or geodes, in the rock. Chert balls or nodules and stringers are common in limestone layers, especially in the Florence Limestone of Permian age, which crops out in the Flint Hills area from Marshall County southward to Cowley and Chautauqua counties.

Most limestones are marine deposits, but some are formed in lakes, in rivers, and on land. Kansas has as spring deposits or as caliche beds many

Figure 26—Stone fenceposts are common in the post-rock country of north-central Kansas, but stacked rocks are also used in fences as in this picture from Chautauqua County.

different varieties of limestones, which are described separately as to occurrence and origin.

Limestone is used in making Portland cement, and the alternating limestone and shale beds of eastern Kansas are a good source of the two chief ingredients of the cement. Portland cement is manufactured in Allen, Montgomery, Neosho, Wilson, and Wyandotte counties. Limestone is used in the construction of roads and railroads, and of fences, as a building stone, as a soil fertilizer, as a source of lime in chemicals, and in many other ways. Chalk, one type of limestone, is an ingredient in paints and polishes.

Limestones Consisting Mainly of Animal Shells

The shells of many animals, those that live either in the sea or in fresh water, consist of calcite and another mineral, aragonite, which is also composed of calcium carbonate. When the animals die their shells are left on the ocean floor, lake bottom, or river bed, where they may accumulate into thick deposits.

Crinoidal limestones.—Crinoids are sea animals having long stems and cuplike bodies that look so much like flowers that they are called sea lilies. The stems break into small, disc-shaped fragments. Some limestones of the Pennsylvanian and Permian rocks of Kansas contain so many of these stem fragments that the term crinoidal limestone describes them well. These are found extensively in eastern Kansas. The Cretaceous Niobrara chalk in a few localities in western Kansas contains beautiful specimens of stemless crinoids, among which both the bodies and the long arms are excellently preserved.

Fusulinid limestones.—One group of the one-celled animals called Foraminifera is known as the fusulinids. These small animals, whose shells look like grains of wheat, were very abundant during the Pennsylvanian and Permian periods, and many of the rocks of those ages are almost solid masses of fusulinid shells. Both limestones and shales contain multitudes of these animal shells.

Reeflike limestones and shell limestones.—Many limestones contain the shells not only of crinoids and fusulinids, but also of corals, brachiopods, clams, oysters, bryozoans, and other forms. Some of the animals lived in colonies, and the remains formed lens-shaped or elongate deposits, which may grow to several hundred feet in thickness and hundreds of miles in length, but those in Kansas are much smaller. Reeflike bodies in eastern and southeastern Kansas were formed by limy mud trapped by leaflike blades of marine algae. (One limestone formation in Labette County has a reeflike structure at least twelve miles long and thirteen feet thick.) Small colonies or groups of fossil shells in some formations measure only a foot in diameter.

Limestones Formed Partly by Chemical Processes

Calcium carbonate is more soluble in water that contains carbon dioxide than in pure water, and when the carbon dioxide is removed for any reason, the calcium carbonate falls out of solution and settles to the bottom. Plants remove carbon dioxide from the water by using it in their food. When the water is heated, evaporated, or merely stirred, the carbon dioxide content is decreased and limestone is deposited.

Algal limestone.—Algae are primitive plants; most seaweeds and pond scums are algae. They may live in sea water or fresh water. Like all plants, they use carbon dioxide to manufacture food, and when the carbon dioxide comes from water containing calcium carbonate, the calcium carbonate may be precipitated. The limestone that results commonly takes on the form of algae or groups of algae and may form irregularly shaped, banded structures. About half of the limestones of the Pennsylvanian and Permian rocks of eastern Kansas are at least partly formed by algae. Particularly good exposures can be seen in Johnson and Brown counties and in the Flint Hills.

Figure 27—Erosion-resistant rock layers cap the chalk formations at Monument Rocks, Gove County.

Oölitic limestones.—Oölites are small rounded particles or grains, so named because they look like fish eggs. Oölites are commonly formed by layers of material, usually calcite, that have been deposited around some tiny particle such as a sand grain or fossil fragment. When the grains formed by this method are more than two millimeters in diameter—about the size of the head of a pin—they are called pisolites.

Many limestones in Kansas, particularly limestones of Pennsylvanian age, contain oölites. They are especially noticeable in Johnson, Miami, Linn, Bourbon, and Labette counties, and near the towns of Independence and Cherryvale in Montgomery County. Rocks of Permian and younger age in Kansas do not contain many oölites; hence most outcrops of oölitic limestone are found in the eastern third of the state. Some of the oölites may be of algal origin.

Figure 28—Erosion takes its toll on Monument Rocks, Gove County.

Chalk.—Chalk is a variety of limestone that probably was formed in either or both of two ways. Part of the limestone is the accumulation of shells of the small, one-celled animals called Foraminifera; the rest of the limestone resulted mainly from chemical precipitation of calcium carbonate. Pure chalk is white, but it may be stained with iron oxide or other impurities. It is a soft, porous rock that crumbles easily.

In the Upper Cretaceous rocks of western Kansas, a chalk and chalky shale formation, the Niobrara, crops out in an irregular belt from Smith and Jewell counties on the northeast to Finney and Logan counties on the southwest. In color the rock is gray to cream but weathers white, yellow, or orange. The average thickness of the entire formation, including pure chalk and chalky shale, is about six hundred feet.

Chalk is abundant in west-central Kansas and is as representative of Kansas geology as the sunflower and the cottonwood tree are of the state's plant life or the meadowlark and buffalo of the animal kingdom. Furthermore, chalk has been a part of the Kansas region not for centuries but for thousands of centuries.

Figure 29—Waconda Spring in Mitchell County, now covered by Glen Elder Reservoir.

Kansas chalk beds are known the world over for the reptilian and other vertebrate fossils found in them, and they are equally famous for the pinnacles, spires, and odd-shaped masses formed by chalk remnants in many localities. Particularly notable are Monument Rocks and Castle Rock in Gove County and the chalk bluffs along Smoky Hill River in Logan, Gove, and Trego counties.

Diatomaceous marl.—This rock is important not because of the calcium carbonate it contains, but because of the diatoms—tiny, one-celled creatures that have characteristics of both plants and animals and have outer shells of silica. When diatoms die, their shells settle to the bottom of the lake or sea and accumulate. In the Ogallala Formation of Wallace and Logan counties there are deposits of white marl containing many shells of these small diatoms. The deposits can be seen plainly from a distance where they crop out along the south side of a valley for about four miles. Outcrops of diatomaceous marl are also found in Meade and Seward counties and elsewhere. The marl can be mined and used for filtering water or other solutions, and as a filler in paints and other products.

Travertine.—Travertine is formed along streams, particularly where there are waterfalls, and around hot or cold springs. Calcium carbonate is deposited because evaporation of the water leaves a solution that is supersaturated with chemical constituents of calcite. Travertine is a banded and more or less compact variety of limestone. A good deposit of travertine in Kansas was once found around Waconda Spring in Mitchell County where the minerals in the spring water gradually built a hill of travertine forty-two feet high and three hundred feet in diameter. The travertine of Waconda Spring, now covered by the waters of Glen Elder Reservoir, was formed chiefly of the mineral aragonite, an unusual form of calcium carbonate (see Figure 29).

Tufa, a porous or cellular variety of travertine, is found near waterfalls. Such deposits have been reported in Riley and Butler counties, and they probably occur in many more places.

Caliche.—Caliche is a type of calcite-cemented sandstone that forms in the soils of dry regions. It is generally impure—clayey, silty, sandy—and fairly soft, although very old caliche may be extremely hard. Some caliche consists only of small nodules, but some caliche occurs as a continuous bed that can be traced for many miles. The pisolitic limestone near the surface of the High Plains in western Kansas is such a bed. This dense limestone (formerly described as algal) has a distinctive structure and can be recognized by its pinkish color, banded appearance, and concentric areas. It was formed after the close of Tertiary time, when the climate was drier than it is now.

Dolomite

Dolomites are fine- to coarse-grained carbonate rocks that are similar to limestones. In the unweathered state, dolomite is generally gray to light colored. On weathering they may become rust-colored because of the oxidation of iron, which occurs commonly as small amounts of pyrite. Dolomites may be formed by the same processes as limestones are formed— for example, by chemical precipitation, or occasionally by breakdown and redeposition of older dolomites. They are also converted from limestone by a process appropriately called dolomitization. This involves the replacement of some calcium by magnesium. The change may take place before or after the rock has been solidified and is caused by the action of sea water, ground water, or hot mineral water.

Among the surface rocks of Kansas, dolomite is found chiefly in three formations in the central and southern part of the state. One formation (the Stone Corral Dolomite) has a maximum thickness of about six feet in Rice County. Another formation (Day Creek Dolomite), found in Clark County, is about two and a half feet thick. These formations were deposited in the enclosed evaporating basin of the Permian sea, evidenced by the presence of much anhydrite, an evaporite deposit in the subsurface rocks. A thick dolomite is also found in the southern Flint Hills. Dolomite is used in most of the ways that limestone may be employed.

Clay

Clay is one of the most common earth substances in Kansas. It is a very fine-grained material that can be molded into shapes and can be heated or baked into hard, resistant forms that have many uses.

The particles in a clay deposit are so small (less than $1/256$ millimeter in diameter) that they cannot be seen without a microscope. Formed by weathering and breaking down of solid rocks, these particles may then be carried to some quiet body of water, such as a lake or pond or the sea, where they settle to the bottom. Clay particles are made of several types of minerals, most of which are small, platy flakes. The deposits formed may have almost any color—white, gray, black, red, yellow, buff, or green. Many clay deposits contain impurities such as sand, calcium carbonate (the chemical compound that forms limestones), and iron minerals.

Clays and shales are used in making bricks, tiles, pottery, chemical ware, furnace linings, and lightweight concrete aggregates. In Kansas, the best deposits of refractory clays—those that can withstand firing at high temperatures—are in rocks of Cretaceous age in Washington, Clay, Cloud, Lincoln, Ottawa, Ellsworth, and other central Kansas counties. Thousands of tons of these Cretaceous clays are used each year in the manufacture of light-colored face bricks, and smaller amounts are used in making pottery.

Bentonite.—This is a clay formed by chemical alteration of volcanic ash. When water is added to bentonite, it forms a milky cloud in the water. Most Kansas bentonites swell, when wet, to less than three times their original volume, but some may swell to as much as fifteen times the original bulk. Some bentonites can be identified by their waxy or soapy appearance. Many deposits are known in western Kansas, the thickest in Phillips County. Very thin layers can be seen at McAllaster Buttes in Logan County, and a thin layer occurs in places above the volcanic ash in the Calvert ash mine in Norton County. Other thin deposits are interbedded with the chalks and chalky shales in western Kansas, and there is a particularly pure bed in Clark County.

Underclay.—This is a clay that occurs under a coal bed or under a coal horizon. This clay, generally characterized by lack of bedding, commonly contains fossilized roots of plants and other carbonaceous material. Underclay is present under many of the coal beds in southeastern Kansas, and in Cherokee County it is utilized in the manufacture of buff brick. Some of the underclays are suitable for firing at high temperatures.

Figure 30—A brick plant near Kanopolis in Ellsworth County; the brick is produced from Cretaceous clays.

Shale

A hardened, compacted clay or silty clay that commonly (but not in every case) breaks along bedding planes is called shale. The particles that make up shale are too small to be seen without a microscope. Many shales have a leaflike bedding and weather into thin slabs or plates, some of which are no thicker than paper. When shales weather they form clays or muds.

Shales and clays are easily eroded, or worn away. Consequently, the best exposures are found beneath ledges of harder, more resistant rocks, such as limestones and sandstones. Most shales are soft enough to be cut with a knife and are either brittle or crumbly. They are usually gray, but black, green, red, or buff shales are common. Many contain nodules of pyrite, selenite (gypsum) crystals, or concretions of various forms, which are described in chapter 3.

Shale and clay together make up about 80 percent of the sedimentary rocks of the earth's crust. In Kansas they are very common. Dark gray to black Cretaceous shales, hundreds of feet thick, are common in the west-central part of the state. Much of the shale found in Pennsylvanian rocks is interbedded with layers of limestone. Shale in eastern Kansas has been used for many years in making bricks. When heated, its color changes to well-known brick-red. Shale is also used mixed with limestone for making Portland cement.

In certain eastern counties, particularly Labette and Neosho, there are several black, platy shales that contain large amounts of organic matter. Some are so rich in this material that thin slivers may be set on fire with a match; they are a form of oil shale.

Some black, very thin bedded shales are often called slate because they have the same color as many slates and because they break into thin, hard, platy sheets. Slate, however, is a metamorphic rock, formed when shale is put under great heat and pressure, and no true slate occurs naturally in Kansas.

Silt and Siltstone

Silt is a common sedimentary rock composed of tiny particles smaller than sand size yet larger than clay size ($\frac{1}{16}$–$\frac{1}{256}$ millimeter). It is found in stream deposits and lake beds, but it occurs chiefly as a wind-blown deposit, called loess, which mantles the High Plains of western Kansas, and in thick deposits along the bluffs of the Missouri River in northeast Kansas. Loess occurs to some extent in many other counties in the state. It is typically a yellowish-buff, porous silt that crops out with steep faces along hillsides and valley walls. Much loess contains white or cream-colored concretions an inch or two in diameter, which are composed of calcium carbonate and have been called *Loess Kindchen* (little children of the loess). Small white shells of snails may also be found in the loess.

Some of the finest and thickest soils in the world are formed in the upper part of thick deposits of loess. As wind moves small particles only, a soil developed in deposits of this kind is free from boulders and pebbles. Loess deposits have been built up by successive dust storms. More than 90 percent of the soil in Thomas, Sherman, Cheyenne, Greeley, Hamilton, Wichita, Scott, Lane, and other western counties consists of the upper part of these loess deposits. In Ford, Grant, Gray, Haskell, Kiowa, Meade, and Stevens counties in southwestern Kansas, loess deposits also are widespread. In short, those dust storms past and present, which we consider as damaging to the state, have helped to give us one of our most valuable resources—a rich, fertile soil.

In northeastern Kansas a very rich soil has developed on the loess, especially in Brown and Doniphan counties and along the bluffs of the Missouri River as far south as Kansas City. This loess is present on the hilltops, on the slopes, and in the valleys. It is thickest in a strip about two miles wide bordering the Missouri River and in Doniphan County in the area of the big bend of the Missouri River. On the river bluffs the loess is sixty to a hundred feet thick; farther from the river it may be no more than five feet thick. This loess of northeastern Kansas is the fine material ground by the advancing ice sheet and deposited on the flood plains by streams coming from the melting glaciers. The material was later worked over by winds. It is thought that most of the loess in northeastern Kansas was laid down more than fifty thousand years ago.

Consolidated or compacted silt is known as siltstone. This rock may be found as thin, slabby beds in many of the Pennsylvanian formations of eastern Kansas. Many siltstones and fine sandstones contain layers rich in tiny flakes of mica, which glitter in the sun. The mica is concentrated along the bedding planes where the rocks break easily.

Figure 31—Palmer's cave, an opening in an outcrop of Dakota Sandstone, in northeastern Ellsworth County.

Plate 1—Castle Rock, a chalk monument in eastern Gove County, was a landmark along the Butterfield Stagecoach route.

Plate 2—Schermerhorn Cave, an opening in Mississippian-age rocks south of Galena, Cherokee County. These rocks are among the oldest at the surface in Kansas.

Plate 3—A waterfall, Cherokee County. The lichen-covered rocks are Mississippian-age limestone.

Plate 4—Pennsylvanian-age sandstone, Chautauqua County.

Plate 5—Osro Falls, along the Caney River in Chautauqua County. Straight lines in the waterfall show natural joints in the underlying limestone.

Plate 6—Controlled burning of Flint Hills grass in the spring.

Plate 7—Flint Hills, Morris County.

Plate 8—Layers of flint in limestone, Riley County.

Plate 9—Glacial debris atop a hill south of Wamego, Wabaunsee County.

Plate 10—Alcove Springs, a stop along the Oregon Trail, in Marshall County. Note watercress in lower left of picture.

Plate 11—Pillsbury Crossing, a waterfall along Deep Creek, east of Manhattan, Riley County.

Plate 12—The Hart House, made of native limestone, in Beloit.

Plate 13—Stone fencepost, Russell County.

Plate 14—Mushroom Rocks State Park, Ellsworth County.

Plate 15—Two pinnacles in the Red Hills of Barber County.

Plate 16—Lines of gypsum cross the Permian-age Red Hills of Clark County.

Plate 17—The Little Basin, a sinkhole about half a mile in diameter, in Clark County. St. Jacob's Well is in the center of the photograph.

Plate 18—Big Basin, a sinkhole that is about a mile in diameter, in western Clark County. This photo shows approximately the northern half of the sink.

Plate 19—Chalk badlands in Logan County.

Plate 20—Chalk makes up a variety of formations at Monument Rocks, Gove County.

Plate 21—Another view of Monument Rocks.

Plate 22—Loess deposits form a rugged topography in Cheyenne County in northwestern Kansas.

Plate 23—Elephant Rock, an eroded outcrop of the Ogallala Formation, in northwestern Decatur County.

Plate 24—A sand dune south of Lakin, Kearny County.

Plate 25—Sphalerite from Cherokee County.

Plate 26—Galena, from the tri-state district.

Plate 27—Opaline sandstone from Rooks County.

Plate 28—Moss opal from Gove County.

Plate 29—Reconstruction of a site in the Garnett locality, of upper Pennsylvanian age, 270 million years ago.

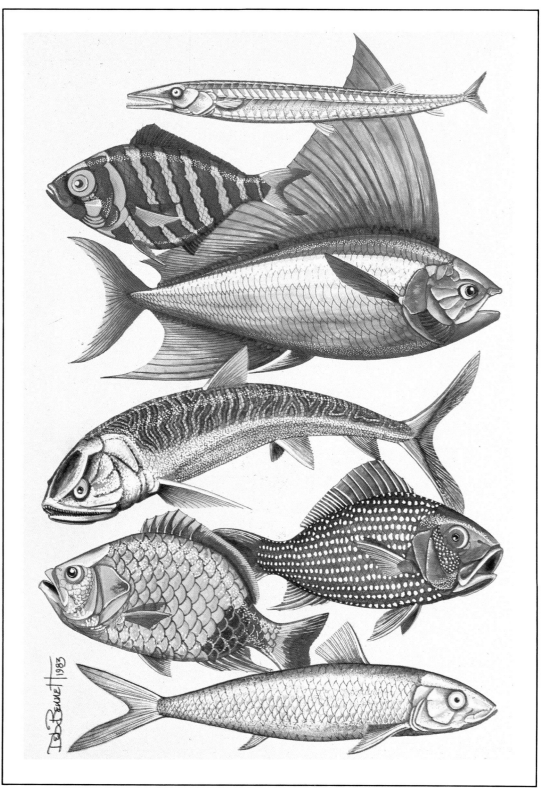

Plate 30—Fishes of the Kansas Cretaceous, about 83 million years ago. From the top down (maximum lengths shown in parentheses): *Leptecodon* (twelve inches); *Omosoma* (two inches); *Bananogmius* (six feet); *Pachyrhizodus* (four feet); *Kansius* (five inches); *Caproberyx* (seven inches); *Apsopelix* (one foot).

Plate 31—A reconstruction of the Rhinoceros Hill site, Wallace County.

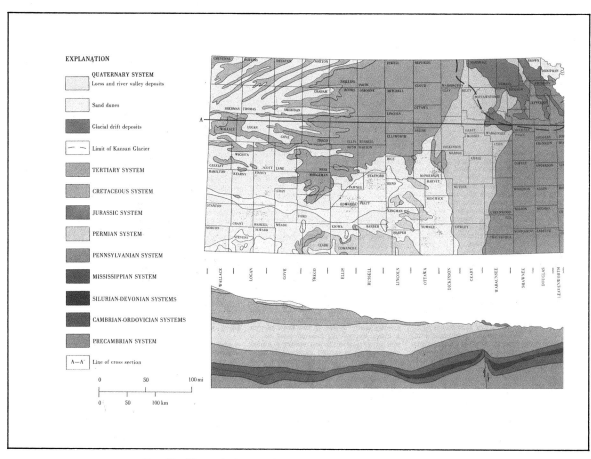

Plate 32—The geologic map shows the ages of the rocks found at the surface of the state. With a few exceptions, formations are progressively younger, moving east to west. The cross section is an approximation of the systems beneath I-70.

Sands and Sandstone

Sands are loose, unconsolidated rocks having particle sizes between those of silt and pebbles ($^1/_{16}$–2 millimeters). When held together by chemical cement or by clay, they are called sandstones. These rocks result from the breaking down or weathering of older rocks and from the transportation and sorting of the rock fragments by moving water or by wind.

Sand.—Sand is found abundantly in Kansas (see Plate 24). Most of it consists of grains of quartz but some of it contains a large amount of feldspar. The sand also contains traces of rare igneous and metamorphic minerals formed outside the state and carried in, along with the other grains, by running water. Sand occurs almost everywhere along the large stream valleys in the state, in regions of old glacial drainage or outwash, particularly near Atchison in Atchison County, or in great deposits of windblown sand in dunes along the Arkansas River in Hamilton, Kearny, Finney, Gray, Ford, Kiowa, Edwards, Pratt, Pawnee, Stafford, Barton, Clark, and Reno counties. Also there is loose sand in parts of the Tertiary Ogallala Formation, along old river deposits in western Kansas.

Sandstones.—Like sand, sandstones consist largely of quartz grains, but they are held together by some natural cement or matrix such as calcium carbonate, iron oxide, or clay, and the rocks can be classified according to the type of cement.

Sandstone occurs interbedded with shale and limestone in the eastern part of the state. In Pottawatomie, Wabaunsee, Nemaha, Brown, Jackson, Riley, Shawnee, Lyon, Greenwood, Cowley, and Chautauqua counties it occurs as channel deposits cutting through shale and limestone. Most of this sandstone is buff or brownish in color, and some is cemented by iron oxide. Sandstone of the Dakota Formation is present in the Smoky Hill region in north-central Kansas in a wide belt extending from Rice and McPherson counties to Washington County. Much of it is cemented by dark brown iron oxide and is so resistant to erosion that it caps steep hills. This Dakota Sandstone also forms the giant concretions at Rock City near Minneapolis and at Mushroom Rocks State Park.

In the same region there are some small areas where the sandstone is cemented by calcite (calcium carbonate) in crystals so large that wide areas of the rock reflect light in a manner known as "luster-mottling." This rock is sometimes called quartzite because it is very hard, but it is not a true quartzite. True quartzites are cemented by a form of quartz commonly called silica. Soft, crumbly sandstones of Cretaceous age form cliffs and box canyons in small areas in Kiowa, Comanche, and Clark counties. Iron oxide cement in this sandstone (the Cheyenne) produces a wide range of colors, including yellow, various bright reds, purple, and brown. Bright red, fine-grained sandstones of Permian age crop out in picturesque canyons in south-central Kansas.

The most common outcropping rock in the Ogallala Formation of Tertiary age in western Kansas is sandstone cemented by very fine grained calcium carbonate. This rock is porous and the particles are poorly sorted. It looks much like concrete and is popularly known as mortar beds. Good outcrops of the mortar beds occur near the town of Cedar Bluffs in Decatur County. Hard, dense, gray-green sandstone is also found in some parts of the Ogallala, especially in southern Phillips County, but also in Graham, Hodgeman, Ness, Norton, Rawlins, Rooks, and Smith counties. This rock has an opal cement, and therefore is called opaline sandstone or ortho-quartzite (Plate 27). It is occasionally used as a building material, and good examples are present in a Hill City park.

Sands and sandstones are used in making glass, as building materials (mainly in concrete), as filters, and for making molds in foundries. The supply along the larger rivers is abundant, although some of the deposits are impure.

Gravel and Conglomerate

The term gravel, used loosely, usually means a rock composed of particles ranging from sand to pebble size or larger (2–64 millimeters). Gravel deposits vary greatly in mineral composition, size, shape, and color. Glacial outwash gravels occur in northeastern Kansas; stream gravels underlie the High Plains in the western part of the state. In the Flint Hills section and in southeastern Kansas there are gravels that consist mainly of just one mineral—chert, or flint, which weathered from Paleozoic limestones. These brown, hard, resistant gravels commonly cap the uplands in Anderson, Cowley, Elk, Greenwood, Lyon, Morris, Wabaunsee, Geary, Riley, Pottawatomie, and Marshall counties.

The gravels from the continental glaciers, and those gravels in the western part of the state that came from the Rocky Mountains, are excellent sources of many rocks and minerals that are not found in place in the outcropping sedimentary rocks of Kansas. Among these are feldspar, agate, clear transparent quartz, native copper, granite, basalt (a dark, fine-grained igneous rock), and other igneous rocks.

Conglomerate is a hardened, generally cemented gravel. Like sand, silt, and clay, it has been formed by the breaking down of older rocks and by later redeposition. Commonly it is found interbedded with layers of sandstone. It also occurs at the base of many Pennsylvanian formations, as for example near the town of Baldwin in Douglas County. Local small areas of hard conglomerate are found in many gravel pits in Tertiary and Pleistocene deposits. Conglomerate and gravel are used in making concrete, in surfacing roads, and as railroad ballast.

Boulder Clay

Boulder clay is an extremely varied deposit consisting, as the name suggests, of particles of all sizes from large boulders to clay. It is a typical product of glacial action and is often called glacial till. Therefore, boulder clay is found only in the northeastern section of Kansas, the only part of the state that has been glaciated. Boulder clay forms low, rounded, rolling hills covered with loess, soil, and vegetation; consequently, there are very few good outcrops. The boulders and pebbles, which have been carried by ice from both local and distant rocks, are of many different types. They include limestone, sandstone, quartzite, granite, basalt, and many others. The quartzite boulders are hard, red rocks; their nearest source is southeastern South Dakota, several hundred miles away. Most of these rocks have been deeply weathered since they were left by the melting of the ice sheet; often they have been weathered so much that a "hard" granite can be crumbled with bare hands. Some of the quartzite, however, is so hard and well preserved that it cannot be broken with a hammer. Upon close examination, some pebbles and boulders are seen to have been scratched and polished from rubbing against other rocks in the ice.

The glacial till conceals the bedrock over much of the glaciated region. This is one of the reasons that northeastern Kansas was overlooked so long in prospecting for oil and gas. However, the thick deposits of boulder clay have formed deep soils, which are especially good for fruit crops. Furthermore, the streams cutting their valleys into the unconsolidated glacial material have produced a type of scenery that is not found in other parts of the state.

Evaporites

Rocks formed by the evaporation of water are known as evaporites. This evaporation may take place either in shallow basins on the land or in the sea; however, the rocks that were laid down under the sea form the thicker and more widespread deposits. Kansas rocks formed in this way include deposits of gypsum, anhydrite, and common salt, or halite.

Sea water contains many salts in solution. These are brought into the oceans by rivers, which are continually wearing down or eroding land surfaces and dissolving the salts. When the sea water evaporates, the salts precipitate and settle to the bottom. The less soluble compounds—those that dissolve less readily in water—are deposited first during the evaporation process. Calcium sulfate, the compound that forms gypsum and anhydrite, is among the least soluble and consequently is one of the first deposited after dolomite. Next in order of solubility and hence in deposition is sodium chloride, or common table salt.

The Permian sea in which evaporites were deposited in Kansas was a shallow arm of the ocean that was shut off from the main body of water by some barrier, perhaps land areas in Oklahoma and Texas. The rate of evaporation was greater than the combined inflow of water from the ocean and from rainfall, and as evaporation continued, the salts of the ocean water became more and more concentrated. Occasionally, more water from the ocean came into the Kansas sea, and this in turn was evaporated. Gradually thick deposits of gypsum, anhydrite, and salt were built up on the sea bottom. These were buried by later deposits of Permian age and then by younger rocks.

Evaporite deposits are described more fully under the separate mineral names gypsum, anhydrite, and halite. They are very common in Kansas, particularly in the central and western parts of the state, and they have many uses.

The evaporites formed on the land are neither so thick nor so common as those formed under the sea. However, at various places in Kansas—especially in northeastern Stafford County and near Jamestown in Cloud County—salt flats occasionally form (see Figure 32). These result from the solution of gypsum or halite by ground water that later evaporated when it reached the surface and left the gypsum and salt deposited on or near the top of the ground. Gypsum deposited in this way looks like dark granular earth and is called gypsite or gypsum dirt.

Figure 32—The Great Salt Marsh, in Stafford County.

Figure 33—Dry Lake, in southwestern Scott County. Water flows into this area, where the ground has subsided slightly, but it cannot escape; after evaporation, a salty crust covers the shore around the lake.

Quartzite

Quartzite is a rock consisting of quartz sandstone so thoroughly cemented with silica that the rock breaks through the grains as easily as around them. It is distinguished from sandstone not only because it breaks through the grains, but also because it cannot be scratched by a knife. Quartzites may be either metamorphic or sedimentary in origin, and the two types are so similar in appearance that in many cases they cannot be distinguished without a microscope. Kansas has small quantities of both types.

Metamorphic quartzites are caused by intense folding of the rock and by solutions from nearby igneous intrusions, or both. The only metamorphic quartzite known in Kansas is located in Woodson County, four miles west and ten and a half miles south of Yates Center, and is probably associated with an intrusion of igneous rock in that area. The quartzite of Woodson County is a thin-bedded, slabby, hard rock of many different colors—green, gray, pink, and black. It crops out in a hillside for hundreds of feet. During the 1870s this location was the site of an unsuccessful silver-mining rush, and many old prospecting pits can still be seen.

Sedimentary quartzite in Kansas is found locally in two Cretaceous formations, the Kiowa Shale and the Dakota Sandstone, where quartzite caps hills and forms hard, resistant ledges in eastern McPherson County and in Kearny County near Hartland. The silica cement was deposited from solution in sea water during or shortly after deposition of the sand. The rock ranges in color from white to brown and light red. The green opal-cemented rock in the Ogallala Formation has been described in the section on sandstones, but it is almost as hard as a true quartzite. As opal is a form of silica, this rock may be considered a special type of quartzite. Typical quartzites, however, are cemented with quartz rather than with opal.

Quartzite boulders are common in the boulder clay of the glaciated area. The rock making up these boulders is red, brownish-red, or purple, and it breaks with a splintery fracture. It is called Sioux quartzite because the ice brought it to Kansas from the area where the Sioux quartzite crops out—southeastern South Dakota, northwestern Iowa, Minnesota, and adjoining states.

Quartzite, because it is so hard and resistant, can be used as a railroad ballast (the crushed rock upon which the tracks are laid) and in the construction of dams. It is used in some places as road material and as building stone.

Asphalt Rock

Asphalt, a solid or nearly solid organic substance composed of carbon and hydrogen, is formed as the lighter parts of petroleum evaporate and the heavy, tarry residue remains behind. Natural asphalt—that made by nature and not in an oil refinery—is found in Kansas in the pores of both limestone and sandstone. It occurs in rocks of Pennsylvanian age in the eastern part of the state and in small amounts in the Cretaceous sandstone of McPherson County. At one time asphalt rock from Linn County was quarried and used in paving roads. Samples of the Linn County asphalt rock contain approximtately 12 percent asphalt. As many porous sandstones and lime-stones do not have asphalt in them, particularly where the rocks crop out and weathering has had a chance to act on them, much exploration must be done in order to find a good deposit of asphalt rock. This is usually done by drilling shallow test holes.

An interesting type of asphalt rock is a Linn County limestone that was once a coral reef. Many of the tiny openings that were the coral cells or cups are partly filled with asphalt.

Jasperoid

Jasperoid is slightly metamorphosed sedimentary rock in which the lead and zinc ores of the tri-state mining area (a district comprising the southeastern corner of Cherokee County, Kansas, and adjacent parts of Missouri and Oklahoma) are commonly found. A gray- to black-mottled chert, coarser grained than ordinary chert, is the cementing material around angular pieces of the original light-colored chert.

Igneous Rocks

Granite

Granite is a coarsely crystalline igneous rock formed by the slow cooling of hot molten rock deep within the earth. All granite contains quartz and feldspar and a small amount of at least one of several other minerals.

The only granite native to Kansas and exposed at the surface crops out on a low hill or ridge along U.S. Highway 75, eight miles south of Yates Center, Woodson County (see Figure 34). The granite consists largely of cream-colored to white feldspar crystals and bluish quartz. It is badly

Figure 34—Igneous rock is rare at the surface in Kansas; this sample is found south of Yates Center, in Woodson County.

weathered, but fresh specimens may be obtained by breaking into the rock. Most of the granite that is seen at this locality is in the form of residual boulders. Probably the granite exposed here does not represent an igneous body that extends to great depth at this locality. Drilling to demonstrate tonnage of the granite revealed peridotite and metamorphosed and unmetamorphosed Pennsylvanian sediments below the surface. The granite has been age-dated as Precambrian, whereas the peridotite is Cretaceous. It is apparently a large boulder that was ripped loose at depth and carried to the surface by the eruptive force of younger igneous rocks.

Tremendous quantities of granite and granitelike rock occur in the subsurface rocks of Kansas. This granite is the basement rock upon which the oldest Paleozoic sediments were deposited, and it is found at depths of from six hundred to eighty-seven hundred feet. It is closest to the surface in Nemaha County.

Figure 35—Most igneous rocks in Kansas have been transported into the state from other areas. This sample of andesite, about twice the size of a softball, is from Cheyenne County, but it was probably washed into Kansas from Wyoming.

Specimens of many varieties of granite and other igneous rocks are found in the boulder clay or glacial till in northeastern Kansas. Cobbles and pebbles of granite are also found in western Kansas, brought there from the Rocky Mountains by prehistoric streams.

Peridotite

Peridotite crops out in a sill-like mass about a mile long and one-fourth mile wide in southern Woodson County. Peridotite is a medium- to coarse-grained, basic igneous rock, which in this area contains phlogopite mica. The rock is altered to a yellowish-gray mass of clay, studded with vermiculite, at the surface. Kimberlite, a form of peridotite, is found in several locations in Riley County. Kimberlite is formed much like a volcano, as igneous rock wells up to the surface from deep underground. Therefore, these rocks, like the granite, are called intrusive igneous rocks because they were forced into other rocks below the surface of the earth. In appearance the kimberlite is a soft, dull gray-green rock and is cut by thin white veins of calcite and magnetite. Kansas peridotite also contains irregular spots of hardened or altered shale. When the hot molten igneous rock came in contact with the Paleozoic shale, or country rock, pieces of shale broke off and fell into the liquid mass. Several kimberlites also contain such minerals as ilmenite and garnet. In some places, such as South Africa, diamonds are formed in kimberlites, although no diamonds have been found in the Kansas kimberlite.

Volcanic Ash

Volcanic ash, or volcanic dust (in some places called silica, although this name is not exactly accurate), consists of tiny glass or congealed lava fragments that have been blown into the atmosphere during the eruptions of volcanoes such as Mount St. Helens. It is a type of extrusive igneous rock; that is, it has been forced out, or extruded, onto the earth's surface. Volcanic ash in Kansas is found in sedimentary deposits of Tertiary and Quaternary age. In some places it is more than twenty feet thick. Under a microscope or a hand lens, ash is seen to contain small, curved pieces of glass that are the broken walls of bubbles of the lava rock that burst from the volcano. Kansas volcanic ash has about the same chemical composition as granite, but in the case of the ash the molten rock cooled so quickly that there was not time for crystals to form. Ash can easily be distinguished from other rocks by its white to bluish-gray color. Its glassy surfaces sparkle in the sun, and its particles do not dissolve in water as do particles of limestone and chalk.

There were no volcanoes in Kansas during Tertiary and Quaternary times so a source outside the state must have been responsible for the large quantities of ash deposited here. Most of the Kansas ash was probably carried in by wind from a volcano in north-central New Mexico; from eruptions in Yellowstone, Wyoming; or from the volcano that produced the

Figure 36—Volcanic ash is mined in this pit near Calvert, Norton County.

Long Valley Caldera in California. Once in Kansas, some of it was carried for short distances by streams and was deposited in quiet ponds, burying pond grasses and snails in the clays beneath. Ash occurs abundantly in central and western Kansas and has been found as far east as Nemaha, Douglas, and Chautauqua counties. In the past it has had many uses: in toothpastes and powders, as abrasives, in cleaning compounds, as glazes for pottery, in filters, and in the manufacture of cement and road asphalt. Today it is mined in Lincoln County and near Calvert in Norton County (see Figure 36).

Basalt

When volcanoes erupt quietly instead of explosively, molten rock pours out in a liquid state of varying thickness, depending on the silica content of the lava. The solidified material formed by cooling of the lava sometimes has a ropy appearance; it is a dark, fine-grained rock called basalt. No basalt native to Kansas occurs at the surface, but many persons have mistaken for such igneous rock some of the ropy-appearing, dark-brown sandstone of the Dakota Formation, as seen on Coronado Heights in Saline County. This structure, which had nothing to do with molten rock, was caused by the chemical deposition of iron oxide in the sand.

Boulders and pebbles of basalt occur in stream deposits in the southwestern part of the state and in the boulder clay and related deposits of the glaciated region.

Meteorites

Almost everyone has seen shooting stars, which are properly called meteorites if they reach the earth's surface. They consist of extraterrestrial rock fragments that have come into the earth's atmosphere. The friction between the rock and the air causes the meteors and surroundings gases to glow from the heat, and it is these hot, glowing objects that we see. Most of them never reach the earth's surface because the intense heat consumes them very quickly by vaporization.

There are two main types of meteorites, one easy to identify, the other difficult. The first type, the iron meteorite, consists mostly of the heavy metals iron and nickel. The other variety, the so-called stony meteorite, consists of heavy minerals and looks very much like volcanic rock. Ordinarily, however, these stony meteorites, unlike true volcanic rocks, contain metallic iron. Meteorite fragments vary from pea size to a mass of about thirty-six tons; most of them weigh less than a hundred pounds. In general, meteorites may be distinguished from other rocks in the following ways: (1) as a rule they are denser than other rocks, (2) in all cases they are solid masses of either iron or stone or both, (3) they have a distinct burned appearance, (4) they are commonly pitted or pockmarked, and (5) most of them will attract a magnet because of the iron they contain.

Over the years, many meteorites have been found in Kansas. Lack of vegetation, predominance of sedimentary rocks, and widespread plowing are factors that have led to record finds.

Mineral Fuels

Coal and Lignite

Coal is a general name applied to black deposits consisting chiefly of carbon compounds derived from plants and plant debris that have been compacted into firm, brittle rocks showing either a dull or shiny luster. There are three main types of coal: anthracite, or hard coal; bituminous, or soft coal; and lignite, a soft, low-grade, impure type. Anthracite, which is not found in Kansas, is a dense, brittle coal with either a shiny or dull luster and a shell-like fracture. It burns with a pale blue, smokeless flame. Bituminous coal, although soft, does not crumble on exposure to air. It breaks into irregularly shaped blocks, has a luster varying from dull to fairly bright, and burns with a yellow flame. Lignite, which contains well preserved plant structures (such as ferns, horsetails, and club mosses), originated in swamps.

Kansas coal is mainly bituminous and is found in the eastern third of the state. It has played an important role in Kansas economy and has been mined in Anderson, Atchison, Bourbon, Brown, Cherokee, Coffey, Crawford, Chautauqua, Doniphan, Douglas, Elk, Franklin, Geary, Jackson, Jefferson, Labette, Leavenworth, Linn, Lyon, Montgomery, Nemaha, Neosho, Osage, Shawnee, Wabaunsee, and Wilson counties. Most of it originated during the Pennsylvanian period, sometimes referred to as the Great Coal Age, but a few beds are in Permian rocks. Originally this coal was probably vegetation in fresh-water swamps similar to those found on parts of the Atlantic coast today. After the plants died and were buried under muds and sands, they began to decay. The first stage in the formation of coal is the development of peat. The passing of geologic time changes peat

Figure 37—Strip-mining coal east of Arma, Crawford County.

into lignite, or brown coal, and eventually into bituminous coal. Had Kansas coal undergone even more heat and pressure than it has, it might have become anthracite.

Brown coal is found in Kansas in younger rocks of Cretaceous age. Small quantities occur in the Dakota Formation in Clay, Cloud, Dickinson, Ellsworth, Ford, Hodgeman, Jewell, Lincoln, Mitchell, Republic, Russell, Washington, and perhaps a few other counties. Lignite coal, which is woody in appearance, is intermediate in quality between peat and bituminous coal. Its water content is as much as 40 percent, and when exposed to air it dries out and crumbles.

Petroleum and Natural Gas

Petroleum is an oily liquid consisting of many compounds of hydrogen and carbon. Most of it is found by drilling to subsurface deposits in rock formations. Rarely, some may be seen oozing from cracks in rocks or floating on the surface of water. Oil seeps have been reported in Smith County, southeastern Kansas, and a few other places where oil coverings have been found on the surface of water in wells and ponds.

Oil, like coal, is formed from the remains of living organisms. It is formed by gradual geothermal "cooking" that transforms the most mobile hydrocarbon chemicals over a few million years. Unlike coal, however, oil usually moves from the source rock to a reservoir rock after its formation. Oil in the reservoir rock occurs in the pore spaces in limestone or dolomite or between sand grains; it is not found in underground pools or lakes. Oil wells are drilled with the hope of finding a porous reservoir rock containing a commercial accumulation of oil. Kansas has produced oil since the 1860s and regularly ranks among the leading states in the nation in terms of oil production and exploration.

Natural gas is associated with oil in most oil fields. In some places—such as the Hugoton Gas Area, one of the world's largest gas fields (Finney, Grant, Hamilton, Haskell, Kearny, Morton, Seward, Stanton, and Stevens counties)—it occurs in huge quantities.

Oil Shale

Oil shale is a compact, laminated sedimentary rock containing a large amount of organic matter that can yield oil when distilled. The organic matter came from partially decomposed algae and animals. Oil shales are an important source of oil in parts of Europe, but deposits in the United States may be regarded as a reserve supply for future needs.

Deposits of black oil shale are found in Labette, Crawford, Bourbon, Douglas, Linn, Neosho, Wallace, and Franklin counties. They occur in Pennsylvanian shales in the eastern counties and in Cretaceous shales in western Kansas.

Figure 38—Drilling for oil in Kearny County, south of Lakin.

Figure 39—Kansas oil production is worth more than two billion dollars annually. This well is south of Ellinwood in Barton County, one of the state's leading oil-producing counties.

3

Minerals and Sedimentary Structures

by Laura Tolsted and Ada Swineford

In several ways, minerals are the key to Kansas geology, for they are the basic components of rocks; all rocks are composed of one or more minerals, just as all molecules are composed of one or more atoms. Some minerals play an important role in the economy of Kansas because they are mined; others may have no monetary value but have an aesthetic appeal. Many of the state's minerals are prized by collectors who admire their unusual forms: selenite, a form of gypsum, makes diamond-shaped crystals; another mineral, opal, is found in particularly beautiful colors in parts of western Kansas; some specimens of galena take on a shiny cube shape. In addition to its beauty, galena is the source of lead ore. It gave rise to the lead-mining industry in southeastern Kansas during the late 1800s and early 1900s. The presence or absence of minerals helps reveal the geologic past to those interested in studying how our land evolved.

Most minerals in the state are sedimentary in origin, as is the case with most Kansas rocks. Salt, a common mineral, was deposited at the bottom of an ancient sea. So was calcite, the primary component of limestone.

What is a mineral? It is a natural, inorganic substance, with a characteristic chemical composition and definite physical properties. "Natural" means that it occurs in nature and is not man-made. "Inorganic" means that the minerals have not been formed directly by any kind of life, either plant or animal. "Characteristic chemical composition" means that no matter where a mineral is found, it contains the same types and quantities of chemical elements. Quartz, for instance, is always composed of the two elements silicon and oxygen, which combine in the proportion of one atom of silicon to two of oxygen to form silicon dioxide, SiO_2. Fool's gold, or pyrite, always consists of iron and sulfur, which form iron sulfide, FeS_2; and sphalerite is always zinc sulfide, ZnS.

The definite physical properties of a mineral depend on its chemical composition and molecular arrangement. It is difficult to describe a mineral without describing its physical properties; the following terms are commonly used to describe minerals.

Color, the easiest physical property to describe, is one of the easiest means of identifying certain pure minerals. For example, gold is always yellow; turquoise is always light blue. Impurities, however, often change the color of minerals.

Luster concerns the character of the light that is reflected by the minerals. Metallic, glassy, earthy, pearly, silky, and similar terms are used to describe luster.

When a mineral is rubbed on a piece of unglazed porcelain, it leaves a streak of finely powdered material. This streak may have a different color from that of the mineral itself and is an excellent check in identifying many minerals. Hematite, the common iron ore, invariably has a red-brown streak.

Some minerals are very soft; others are very hard. The degree of hardness is an aid in identifying the minerals. Diamonds are harder than quartz and will therefore scratch quartz; quartz will scratch calcite; calcite will scratch gypsum; and so on. To help identify minerals, geologists have assigned numbers to the hardness of several minerals. In this hardness scale, the softer minerals are assigned a low number and the harder minerals a higher number.

In the field, an easy way of estimating the hardness of a mineral is by trying to scratch it with common objects such as a fingernail, which has a hardness of 2.5, or a pocketknife blade, with a hardness of 5.5. Glass has a hardness of slightly less than 6 and will scratch most minerals. To test a mineral for hardness, try to scratch it with one of these common objects. Minerals with a hardness of 6 or more will easily scratch a piece of glass. A sample such as calcite is too soft to scratch glass but is hard enough to scratch a fingernail. Therefore it has a hardness between 6 and 2.5. Hardness is another clue in identifying minerals, and in this book the hardness for each mineral is listed alongside its name.

If light passes through a small piece of a mineral so that objects can be seen through it, then the mineral is said to be transparent. If no light passes through and nothing can be seen through a small piece, the mineral is called opaque. Minerals that are neither opaque nor transparent, those through which light passes but through which no objects can be seen, are said to be translucent.

Minerals always take on definite shapes, geometric forms, which may not be visible to the naked eye. The forms, which depend on the arrangement of the atoms and molecules of the mineral, are called crystals. Unfortunately, large and visible crystals form only sporadically, mainly because growing crystals compete for space and crowd each other. Perfect crystals generally are found where they project or grow slowly in open space, such as in fractures or veins. When good crystals are found, they are extremely valuable in identifying minerals. Some crystals are flat and are described as tabular; some are long and thin, like needles or fibers, and are described as either needlelike or fibrous; still others are shaped like pyramids, prisms, cubes, and many other forms or combinations of forms. If the crystal outlines cannot be distinguished and the mineral is composed of compact material with indefinite form, the structure is called massive. Crystals may be grouped into a globular shape that looks like a cluster of grapes. A few minerals seemingly have no crystal structure at all.

When struck a hard blow, some minerals break only along certain planes; other minerals break just as easily in one direction as in another. When a mineral has a tendency to break along certain planes, it is said to have cleavage, which results from the arrangement of the bonds between different molecules and atoms. Minerals may have only one plane of weakness or cleavage, or they may have two, three, or more. The second type of breaking, that which is not determined by any arrangement of molecules, is called fracture, and this also varies among different minerals. Various types of fracture are described as smooth, uneven, ragged, and conchoidal, or shell-like.

Minerals

The following list of minerals has been divided into several groups, based primarily on the chemical composition of the mineral. For example, any mineral that is found in nature in its native state—that is, not combined with any other element—is called a native element. Gold and silver are two well-known native minerals, but neither is found in Kansas.

All other minerals are combinations of elements. Sulfide minerals, which are fairly common in Kansas, are formed by the direct union of an element with sulfur. The combination of zinc and sulfur, for example, produces the sulfide mineral called sphalerite (ZnS). Another common type of mineral in Kansas is silicates, which are complex compounds composed of silicon, oxygen, and one or more metals. One form of mica, the mineral muscovite, is a silicate that is composed of silicon, oxygen, aluminum, and potassium ions.

All of the common minerals in Kansas fall into seven classes of minerals. The following list of minerals shows their classification, where they are found, what they look like, and other information.

Native Elements

There are only about twenty native minerals, those elements occurring in nature in their native state, not combined with other minerals. Sulfur is the only native mineral common in Kansas.

Sulfur (Hardness 1½–2½).—Sulfur (S) occurs as irregular masses, as earthy coatings on other substances, and as fine crystals. It is bright yellow, so soft that it can be scratched by a fingernail, and burns with a blue flame.

Sulfur in Kansas has been reported on the surfaces of coal dumps as slender, needle-shaped crystals resulting from the decomposition of pyrite. Small quantities of the earthy variety are present in many Kansas rocks that contain pyrite. Most of it is impure and mixed with clay and limonite.

Sulfides

Nearly all sulfide minerals are formed by the direct union of atoms of an element with atoms of sulfur. For example, the combination of lead and sulfur forms a sulfide mineral called galena. Many of the sulfide minerals are valuable ores, such as galena and sphalerite—the sulfide minerals that produce lead and zinc ore. In Kansas, many of the sulfide minerals are found in the southeastern corner of the state.

Galena (Hardness 2½).—Galena, the principal ore of lead, is composed of lead sulfide (PbS). It is found in metallic- to lead-gray, cube-shaped crystals that break into cubic, right-angled fragments. Some galena crystals are very large. Galena is heavy, has a metallic luster on fresh surfaces, has a gray-black streak, and is so soft that it will mark on paper.

Galena was once mined in the tri-state district, which was the most important lead- and zinc-producing area in the world in the early part of this century (see Plate 26). In the late 1800s, there were hundreds of small lead and zinc mines in Cherokee County. Today these mines are closed. However, galena is still found with sphalerite, chalcopyrite, cerussite, dolomite, calcite, quartz, barite, and other minerals, especially at old mine dump sites. Galena is also found near Pleasanton, Linn County, and has been reported from Chautauqua, Douglas, Elk, and Sumner counties, and in rock fragments brought to the surface during drilling for oil in many other counties.

Sphalerite (Hardness 3½–4).—Sphalerite, also called zinc blende, blende, black-jack, and mock lead, is composed of zinc sulfide (ZnS) and is the most important ore of zinc. Pure sphalerite is nearly colorless, but it is commonly brown, yellow, black, or dark red because of impurities. It has a white to dark brown streak, always much lighter than the color of the specimen. As a rule the mineral crystals are shaped like triangular pyramids, with three sides and a base; because it has good cleavage in six directions, sphalerite will break into twelve-sided blocks. It has a brilliant, resinous or almost metallic, luster and can be scratched by a knife.

Some sphalerite is found as massive deposits varying from coarse- to fine-grained. In warm hydrochloric acid, powdered sphalerite breaks down and forms hydrogen sulfide, which has a decidedly unpleasant odor, something like the smell of a rotten egg. Sphalerite is easily identified by its cleavage and its resinous luster.

The best specimens of sphalerite found in Kansas are from the lead and zinc mines of Cherokee County (see Plate 25). Sphalerite is also found as small crystals in clay-ironstone concretions in the Pennsylvanian shales of eastern Kansas.

Chalcopyrite (Hardness 3½–4).—An important ore of copper where it occurs in abundance, chalcopyrite is a sulfide of copper and iron ($CuFeS_2$). It is a brassy yellow mineral that makes a greenish-black streak and has a metallic luster. It is brittle, may be tarnished, and can be scratched by a knife. It occurs normally as four-sided pyramidlike crystals, but the crystals are usually poorly formed when the mineral occurs as massive sulfide ore. Chalcopyrite is very similar in appearance to pyrite, but they can be distinguished because each has a characteristic color and hardness.

Chalcopyrite occurs with lead and zinc ores. Although chalcopyrite is mined throughout the world as a copper ore, commercial quantities of the mineral are not known in Kansas.

Greenockite (Hardness 3–3½).—Greenockite, a rare mineral, is composed of cadmium sulfide (CdS). It has a yellow color, resinous to earthy luster, and cannot be scratched by a fingernail. Thin films of greenockite sometimes coat sphalerite and other minerals in the lead and zinc mines of Cherokee County.

Pyrite (Hardness 6–6½).—Pyrite (iron sulfide, FeS_2) is a pale, brassy-yellow, opaque mineral that is brittle and has a metallic luster. It makes a black streak and is so hard that it can scratch glass. Most pyrite crystals are cube-shaped (like galena), but they also occur in other forms such as octahedrons. Pyrite is also found as granular masses, as cones and globules, and as nodules in shale, limestone, and sandstone. It is called fool's gold because its color is yellow, like gold; but pyrite is brittle, has a greenish tinge, and tarnishes, whereas gold is softer, leaves a yellow streak instead of a black one, and does not tarnish easily.

In Kansas, pyrite occurs in rocks of all ages, but it is especially abundant in coal deposits and in areas where lead and zinc occur. It is also found with gypsum in the dark shales. For a few years it was produced as a by-product of coal at West Mineral southwest of Pittsburg and was used in making sulfuric acid.

Marcasite (Hardness 6–6½).—Marcasite, sometimes called white iron pyrite, is a mineral composed, like pyrite, of iron sulfide (FeS_2). Marcasite is a secondary mineral—it forms by chemical alteration of a primary mineral such as chalcopyrite. On fresh surfaces it is pale yellow to almost white and has a bright metallic luster; it tarnishes to a yellowish or brownish color and makes a black streak. It is a brittle mineral that cannot be scratched by a knife. The thin, flat, tabular crystals, when joined in groups, are called cockscombs. When combined into balls or nodules, or into more complicated groups, they are marcasite rosettes. The mineral can be distinguished from pyrite by its crystal form. Marcasite weathers readily to form secondary minerals such as limonite and melanterite.

In Kansas, marcasite occurs as concretions in coal, shale, and limestone. Well-developed crystals have been taken from the lead and zinc mines in Cherokee County and can be found in all of the coal mines in southeastern Kansas.

Oxides

Oxide minerals are those natural compounds in which oxygen is combined with one or more metals. An example of an oxide mineral found in Kansas is hematite (Fe_2O_3), a combination of molecules of iron and oxygen. The oxide minerals are usually harder than any other class of mineral except for the silicates, and they are heavier than the other classes except for sulfides.

Hematite (Hardness $5^1/_2$–$6^1/_2$).—Hematite is a compound of iron and oxygen (Fe_2O_3) that may be either red and earthy, or black with a dull or metallic luster. Both types have a red-brown streak by which the mineral is readily identified. The earthy variety marks paper easily.

Pure hematite is rare in the surface rocks of Kansas. Most hematite in the state is of the red variety and is found scattered in clays and shales. It is the cementing material in red sandstones. Also, small patches of impure hematite mixed with beds of hematite sand are found in the Dakota Formation in eastern Russell County and in Lincoln County near Juniata.

Hematite was once the chief source of iron ore in other parts of the United States, such as northern Minnesota, Michigan, Wisconsin, Pennsylvania, and Alabama.

Ilmenite (Hardness $5^1/_2$–6).—Ilmenite, named for the Ilmen Mountains in the Soviet Union, is an ironlike mineral composed of iron, titanium, and oxygen ($FeTiO_3$). It makes a black to brownish-black streak and cannot be scratched by a knife. Most large specimens of ilmenite are dense, granular masses, but the mineral may occur as platy crystals and as grains in sand. Specimens of the massive variety of ilmenite have been found in the kimberlite near Stockdale in Riley County. In New York State, ilmenite is mined as an ore for its titanium content. Some of the largest deposits of ilmenite occur as beach sands in many parts of the world.

Pyrolusite (Hardness 1–2) and *Psilomelane* (Hardness 5–6).—Pyrolusite and psilomelane are both oxides of manganese (MnO_2), although psilomelane contains varying amounts of other elements. Pyrolusite is a black mineral that is so soft that it will easily make a black streak on paper. It usually occurs in radiating fibers or as treelike patterns (dendrites) on rock surfaces and in the moss agate of Wallace, Trego, and Logan counties. Psilomelane, also a black mineral that makes a very dark brown to black streak, is much harder than pyrolusite—it cannot be scratched by a knife. An earthy form of psilomelane, however, is known as wad, and is soft enough to soil the fingers. Wad forms the coating around pebbles in some gravel deposits, and it also occurs as soft black lumps in gravels and in some soils in southwestern Kansas. Pyrolusite and psilomelane are often found together in the same deposit. Both are sources of manganese ore.

Magnetite (Hardness 6).—Magnetite (iron oxide, Fe_3O_4) is so named because it is readily attracted by a weak magnet and because some magnetite specimens called lodestones are in themselves magnets. The mineral is black, has metallic luster, and makes a black streak. It is so hard that it cannot be scratched with a knife. It is found as granular masses, but, especially in igneous rocks, it occurs as individual crystals, most of which

Figure 40—Salt sample from central Kansas.

Figure 41—Salt mine near Kanopolis, Ellsworth County.

have eight triangular faces and are called octahedrons. Magnetite is an important ore of iron.

Kansas magnetite is found in the kimberlite near Bala in Riley County, where it occurs as tiny, black, shining octahedrons imbedded in the rock. Occasional grains of magnetite may be found in many river sands.

Limonite (Hardness varies).—Limonite, a compound of iron, oxygen, and water [FeO(OH) • nH_2O + Fe_2O_3 • nH_2O], is a yellow-brown to dark brown or black, seemingly noncrystalline mineral. It is formed by the alteration of other minerals that contain iron. Limonite has a characteristic yellow-brown streak, but its hardness depends on the form in which it occurs. The yellow-brown, earthy form of limonite, really a mixture of limonite and clay, called yellow ochre, is so soft that it easily leaves a mark on paper. The dark brown to black variety (bog iron ore) is so hard that it cannot be scratched by a knife. Small quantities of limonite give a yellowish or buff color to most sandstones and to many clays, shales, and limestones. As a scum on quiet water it may be mistaken for oil. It was once an iron ore of minor importance in some states, but Kansas does not have commercial deposits.

Much of the limonite in Kansas is in the form of concretions (particularly in the Dakota Formation) and in the form of impurities in sedimentary rocks. Limonite, which has taken the place of and has kept the crystal form of pyrite (in other words, a pseudomorph after pyrite), has been reported from northeast of Lincolnville in Marion County.

Halides

Halides are compounds that are characterized by atoms of a chemical group called halogens. Halogens include the elements chlorine, iodine, bromine, and fluorine. Because of their chemical makeup, halides are usually soft minerals that have moderate to high boiling points. The only common halide mineral in Kansas is halite, known more commonly as salt.

Halite (Hardness 2½).—Halite, common table salt, is composed of sodium chloride (NaCl). Most of its crystals are transparent, colorless cubes (see Figure 40), but various impurities in the salt may give halite a brilliant red, blue, or yellow color. Normally halite has three good cleavages at right angles to each other, so broken fragments also may be very nearly cube-shaped. Some red halite has a fibrous or columnar structure. Halite is easy to identify because it has a salty taste and because it dissolves rapidly in water.

Salt in Kansas is found in thick beds in Permian rocks deep in the ground. It does not form outcrops because rain and ground water dissolve salt from the surface exposures. Most of the salt is found in the Hutchinson salt bed, which underlies about thirty-seven thousand square miles in central Kansas. The formation is an average of about two hundred and fifty feet thick and contains a staggering thirteen trillion tons of salt, enough to make a block measuring ten miles on each side. It is mined at Hutchinson, Kanopolis, and Lyons and is used in chemical industries, in meat packing, for livestock, and as common table salt. Each year Kansas produces over a million tons of salt.

Carbonates

This group includes some of the most common minerals in Kansas, such as calcite and dolomite. The carbonate minerals are distinguished by a complex chemical makeup that includes an element combined with atoms of carbon and oxygen. Calcite, for example, is composed of atoms of calcium, carbon, and oxygen.

Calcite (Hardness 3).—Calcite (calcium carbonate, $CaCO_3$) is the primary constituent of limestone and is therefore one of the most common minerals in Kansas. Generally it is white or colorless, but it may be tinted gray, red, green, or blue. It occurs in many varieties of crystal forms (more than three hundred have been described). It may also be granular, coarse to fine, or even so fine grained that it has an earthy appearance. Calcite can be scratched by a knife, but not by a fingernail, and it fizzes freely in cold, dilute hydrochloric acid. If a large piece of calcite is shattered with a hammer it breaks into small rhomb-shaped blocks because it has perfect cleavage in three directions.

Besides being the mineral that forms limestone, calcite occurs as a common cementing material in many Kansas sandstones. It is found in calcareous shales and clays, and as veins in the igneous rocks of Riley County. In the Cretaceous Niobrara Chalk and other Cretaceous rocks, it has formed fairly large veins. Calcite is an important part of many concretions; brown calcite and colorless-to-yellow calcite crystals are common in the septarian concretions of the Pierre Shale in Wallace and other counties. Tiny calcite crystals form the lining of geodes in certain limestones and shales, and they coat the insides of many fossil shells. Among the finest calcite crystals in Kansas are those from the lead and zinc mines of Cherokee County; most of these are pale yellow, and some of them are very large. The most important use of calcite is in the manufacture of cements, limes for mortar, in the chemical industry, and in fertilizers.

Siderite (Hardness 3½–4).—Siderite is a common mineral composed of iron carbonate ($FeCO_3$). It is light to dark brown, and some of it occurs as rhomb-shaped crystals with curved faces (like dolomite). Most siderite, however, is granular or earthy. The mineral can be scratched by a knife. It fizzes in hot hydrochloric acid but reacts more slowly in cold acid. Weathered surfaces change to limonite and turn dark brown.

Most siderite in Kansas is in the impure form called clay ironstone. This is a mixture of siderite with limonite, clay, and silt, forming small nodules or whole beds in clays, shales, and sandstones.

Smithsonite (Hardness 5).—Smithsonite (a secondary zinc carbonate, $ZnCO_3$) is commonly brown in color, but it may be green, blue, pink, or white. Although it does occur as rough, curved, rhomb-shaped crystals, its occurrence as rounded, globular forms or as honeycomb masses is more common. Smithsonite is harder than most carbonate minerals, but it can be scratched by a knife. It fizzes in dilute, cold hydrochloric acid.

Smithsonite is common in the near-surface parts of the zinc deposits in easternmost Cherokee County, where it has formed as the result of the action of carbonated water on sphalerite. Smithsonite is a zinc ore, and was named in honor of the Englishman James Smithson, who also supplied funds for the founding of the Smithsonian Institution.

Dolomite (Hardness $3^{1}/_{2}$–4).—The mineral dolomite is composed of calcium magnesium carbonate $[CaMg(CO_3)_2]$ and is closely related to calcite. In large masses the mineral forms the rock called dolomite. It may be white, gray, greenish-gray, brown, or pink, and it has a glassy to pearly luster. It occurs in coarse- to fine-grained granular masses and in crystals. Most dolomite crystals are rhomb-shaped, like calcite cleavage blocks, but (unlike most other minerals) the crystal faces are typically curved. Dolomite is slightly harder than calcite, although it can easily be scratched by a knife. It will not fizz readily in dilute, cold hydrochloric acid unless first ground to a powder.

Curved white crystals of dolomite are common in lead and zinc mines of Cherokee County, where they occur with sphalerite, chalcopyrite, galena, and several other minerals. Dolomite crystals have also been found in Pennsylvanian limestones in the Ross quarry near Ottawa in Franklin County, about two miles north of Williamsburg in Franklin County, and three miles north of Garnett in Anderson County. They also are found in the rock dolomite formations and certain red and green shales of McPherson, Rice, Reno, Kingman, and Clark counties.

Aragonite (Hardness $3^{1}/_{2}$–4).—Aragonite has the same chemical composition as calcite ($CaCO_3$), but it has poorer cleavage and a different crystal form than calcite. Aragonite crystals commonly occur as radiating groups of fibrous or needlelike shapes. Like calcite, aragonite fizzes and dissolves readily in cold, dilute hydrochloric acid and can be scratched with a knife. This mineral—which is colorless to white, gray, yellow, green, brown, and violet—is ordinarily found as a vein mineral, in cave deposits, and as the pearly layer of many types of shells.

Aragonite is much less common than calcite because it changes easily to calcite without altering its external shape. It is difficult to identify in the field. The mineral has been reported from several areas in Kansas: as nodules in a clay deposit in northern McPherson County and in a sand pit two miles southeast of McPherson; as veinlets cutting country rock at Silver City in southern Woodson County; as small crystals in vugs, or cavities, in the limestone of the Ross quarry near Ottawa in Franklin County; and in many concretions in the Cretaceous shales of western Kansas. This mineral is named for the region of Aragon in Spain.

Cerussite (Hardness 3–3½).—Cerussite (lead carbonate, $PbCO_3$) occurs as granular masses and as platy crystals, which commonly cross each other to form a latticelike effect. Cerussite has a brilliant, glassy luster and is colorless or white. It fizzes slightly in cold, dilute hydrochloric acid and is very heavy for a nonmetallic mineral.

In Kansas, small amounts of cerussite are occasionally found as a result of the chemical change of galena (lead sulfide) in the near-surface parts of the lead deposits in easternmost Cherokee County.

Malachite (Hardness 3½–4).—Malachite, a copper ore, is a bright-green copper carbonate mineral having the composition $Cu_2CO_3(OH)_2$. It has a dull to glassy luster and a light green streak. The mineral fizzes in cold, dilute hydrochloric acid and can be scratched by a knife.

Malachite occurs in Sedgwick, Sumner, and Harper counties, where it is associated with copper mineralization in the Permian shales and carbonate rocks. It occurs as tiny, brilliant green specks in some thin dolomite beds near the top of the Wellington Shale and in a few other Permian rocks. It may be found, among other places, at the bridge crossing the Ninnescah River two miles south of Milan, Sumner County. It also occurs in southeastern Kansas.

Sulfates

Another common group of Kansas minerals is the sulfates. These minerals consist of an element combined with atoms of sulfur and oxygen. One of the widespread sulfate minerals is gypsum, which is composed of atoms of calcium combined with atoms of sulfur and oxygen.

Barite (Hardness 3–3½).—Barite (barium sulfate, $BaSO_4$) is a common mineral in Kansas, but it is not found in large quantities. Because of its high density, it is sometimes called heavy spar. It occurs as flat, tabular crystals, either singly or in groups, and it also occurs in granular or earthy forms. The individual crystals are transparent to opaque and have a glassy luster and perfect cleavage in two directions. Barite is usually colorless or white but may be light shades of blue, yellow, or red. It can be scratched with a knife but not with a fingernail. In appearance, it resembles gypsum, calcite, or celestite. However, aside from its relatively heavy weight, it can be distinguished from gypsum by its greater hardness and from calcite because it does not fizz in hydrochloric acid. A flame test is the best means of distinguishing between barite and celestite. If powdered barite is heated on a clean platinum wire in a Bunsen burner, the flame will become green, but celestite will turn the flame bright red.

Barite has been found in Kansas in some of the Pennsylvanian and Permian limestones, especially in Brown, Anderson, Franklin, and Chase counties; in septarian concretions of the Cretaceous Pierre Shale in Logan and Wallace counties; in petrified wood; and occasionally in the lead and zinc mines of Cherokee County. It occurs in veins a few millimeters thick in the Niobrara Chalk in north-central and northwestern Kansas. It also occurs as a cementing material between sand grains in peculiar roselike concretions (called desert roses or petrified walnuts) in certain sandstones of the Cretaceous Kiowa Shale. These barite roses and walnuts are abundant in an area near Bavaria in Saline County.

Barite is used in paint pigments, as a filler in paper and cloth, in making glazes for pottery, and in the refining of sugar. It has not been found in commercial quantities in Kansas but has been mined in Missouri and Arkansas.

Celestite (Hardness 3–3½).—Celestite (strontium sulfate, $SrSO_4$) is similar to barite in appearance, in geologic occurrence, and in crystal form. It has a glassy luster and its crystals are colorless, white, or a faint blue or red. This mineral is also found in Kansas as radiating pink fibers, as vein fillings, and as scattered particles. Celestite cannot be scratched by a fingernail. It differs from barite in its lighter weight and in its property of coloring a flame red.

Celestite has been found in solid blue or pink crystals and as pink to white radiating fibers at Kanopolis dam near the water's edge below the spillway outlet. It has also been found as pink crystals and as veins in Brown County north and west of Morrill and in Chase and other counties. It has been found in the weathered zone at the top of Permian rocks below Cretaceous sands and shales.

Anhydrite (Hardness 3–3½).—Anhydrite, which is composed of calcium sulfate ($CaSO_4$), constitutes one of the three main evaporite deposits, the other two being gypsum and halite. It occurs commonly as light gray, crystalline masses, although some anhydrite has a fibrous structure. It may occur as individual crystals in other rocks, particularly in dolomite. It has a glassy luster and is translucent. It is harder and heavier than gypsum: although it can be scratched easily with a knife, it cannot be scratched with a fingernail. Anhydrite may change into gypsum if water is added, which is common in near-surface exposures. Fine-grained dolomite and anhydrite look somewhat similar but can be distinguished from one another because cold hydrochloric (muriatic) acid will not act on anhydrite.

Kansas anhydrite is found in Permian-age deposits associated with beds of gypsum, dolomite, and red silt. With gypsum, it caps many of the Red Hills of Barber and other counties.

Gypsum (Hardness 2).—Gypsum is calcium sulfate containing water ($CaSO_4 \cdot 2H_2O$). The same chemical compound without water is anhydrite, a quite different mineral. Gypsum is a common mineral that is widely distributed in the sedimentary rocks of Kansas, in the form of thick beds, well-formed single crystals, and joint or crack fillings. It is colorless to white or light gray, rarely bright red, and is so soft that it can be scratched by a fingernail.

Three varieties of gypsum are recognized. The type of most interest to collectors is the coarsely crystalline, transparent variety called selenite. It consists of flat, diamond-shaped crystals (see Figure 42) having such perfect cleavage that they can be split into thin sheets. Selenite is common in dark shales, such as the Kiowa, Carlile, and Pierre shales of Cretaceous age in western Kansas. Most selenite crystals found on weathered shale slopes have irregular or etched surfaces, but fresh, clear crystals can be uncovered by careful digging into the hillside. Occasionally one finds specimens consisting

Figure 42—Selenite, a variety of gypsum.

Figure 43—Gypsum plant at Blue Rapids, in southern Marshall County

of two crystals grown together into what is known as a fish-tail twin. In some places a network of selenite crystals is found in thin joint fillings, and some of the crystals may be grown together in a pattern called gypsum flowers. Small quantities of bright red selenite are found in some of the stream banks on the outskirts of Wichita in Sedgwick County.

Satin spar is another variety of gypsum. It is white or pink, fibrous, and has a silky luster. It is found as thin layers in beds of rock gypsum and in certain shales. The third recognized variety is massive or rock gypsum. It is coarsely to finely granular, white to gray, and contains various amounts of impurities. A good outcrop of rock gypsum can be seen about ten miles west of Medicine Lodge along U.S. Highway 160. Alabaster, which is not known to occur in Kansas, is a very fine grained type of massive gypsum. Gypsite, or gypsum dirt, is formed in the soil or in shallow lakes. Consequently it is a sandy or earthy deposit, although it contains a large amount of the mineral gypsum. It is found in Clay, Saline, Dickinson, Marion, Harvey, and Sedgwick counties.

Large quantities of rock gypsum are mined in Barber and Marshall counties and large deposits are also found in Permian rocks in Saline, Dickinson, and Comanche counties. It is used in making plaster of Paris, Portland cement, various wall plasters and mortars, wallboard, and as a fertilizer.

Goslarite (Hardness 2).—Goslarite ($ZnSO_4 \cdot 7H_2O$), which is zinc sulfate containing water, is formed by chemical action on sphalerite. It is found occasionally in the tri-state area as long, slender, needlelike crystals. Goslarite not uncommonly develops on mine walls. It has a white, reddish, or yellowish color.

Silicates

More than 90 percent of the rock-forming minerals are silicates, compounds containing silicon and oxygen as quartz or combined with one or more metals in more complex molecules. These minerals make up about 95 percent of the earth's crust. Some silicates, such as quartz and feldspar, are especially common.

Garnet (Hardness $6\frac{1}{2}$–$7\frac{1}{2}$).—Garnets are a group of minerals whose crystals have many faces, all of about equal size. They have a glassy luster and are hard enough to scratch window glass. Most garnets are red to brown, but some are black, green, or colorless. In chemical composition they are silicates of calcium, magnesium, iron, manganese, aluminum, and chromium in various combinations, with the aluminum silicate varieties predominating.

Small red and brown garnets occur in the kimberlite outcrop near Stockdale in Riley County, and they may be found in the bed of the small stream that cuts across this outcrop. Garnets have also been found in other Riley County kimberlites and in the streams flowing near the kimberlites.

Hemimorphite (Hardness $4\frac{1}{2}$–5).—Hemimorphite, sometimes called calamine, is a silicate of zinc containing water. Its chemical formula can be written $Zn_4Si_2O_7(OH)_2 \cdot H_2O$. It is a white mineral found in radiating crystal groups and in globular forms. It can be scratched by a knife. Hemimorphite usually occurs with zinc ores, and in Kansas it is fairly common in the upper parts of the sphalerite deposits in Cherokee County. In some parts of the world, hemimorphite is mined for zinc ore, although not in Kansas.

Mica (Hardness $1\frac{1}{2}$–3).—Mica is the name of a group of several minerals that are unusual because they split into thin, flat, flexible sheets. These minerals split this way because they have one perfect cleavage. They are composed of aluminum silicates of several elements. Muscovite, or common white mica, is transparent and colorless. In some igneous rocks (outside Kansas) it occurs in crystals several feet wide, large enough that they were once used as windows in coal stoves. In Kansas it is usually seen as tiny, flat, shining flakes in sandstones, siltstones, and shales and as small crystals in boulders of metamorphic and igneous rocks. In some sands of the Ogallala Formation, muscovite has weathered to resemble gold flakes. Muscovite was so named because it was used as a substitute for glass in old Russia (Muscovy).

Biotite (black mica), rarer in Kansas than muscovite, may be seen in some of the Tertiary and Quaternary sands. The color of biotite is caused by iron. Phlogopite and vermiculite mica are yellowish brown, have a copperlike luster on the cleavage surfaces, and often are mistaken for flakes of gold. Phlogopite is found in the kimberlites of Riley County and near Silver City in Woodson County.

Quartz (Hardness 7).—Quartz, the most common of all minerals, is composed of silicon and oxygen (silica, SiO_2) and is found in many different varieties. When pure it is colorless, but it also assumes various shades of yellow, pink, purple, brown, green, blue, or gray. It has no good cleavage and has a glassy to greasy luster. One of the hardest of the common minerals, it will easily scratch window glass. In fact, quartz can be distinguished from calcite, another extremely common rock, by its hardness. A knife will scratch calcite but not quartz.

There are two primary types of quartz, the coarsely crystalline and the fine or cryptocrystalline forms. The crystals of the first type are six-sided prisms with pyramids capping one or both ends. Well-formed, colorless quartz crystals of this type are found in geodes and as linings on the inside of fossils in many parts of the state, particularly in the Lone Star quarry at Bonner Springs in Wyandotte County and in an area of eastern Chase County. Quartz crystals with a bluish cast are found in the igneous rock south of Yates Center in Woodson County. Nearly all sands and sandstones are composed of tiny, worn particles of crystalline quartz.

The second primary type of quartz is called cryptocrystalline because the crystals are so small they can only be seen with a microscope. One of the best-known varieties of this group is flint or chert, which is common in many Kansas limestones as nodules or beds. Chert is opaque and is dull gray, brown, or black. It breaks with a shell-like fracture, and the edges of the broken pieces are sharp. The famous ''chat mountains'' at the lead and zinc mines of Cherokee County consist almost entirely of chert fragments. Chalcedony is a cryptocrystalline quartz with a waxy luster that forms banded layers or globular masses. Agate is a many-colored, banded form of chalcedony that has been deposited in cavities or in veins and is used for ornamental purposes. Beautiful agates, doubtless from the Lake Superior region, have been found in the glaciated district in Kansas.

Opal (Hardness 5–6).—Opal consists of silicon dioxide, like quartz, plus an indefinite amount of water ($SiO_2 \cdot nH_2O$). It never forms as crystals but probably is deposited as a jellylike substance that later hardens. The mineral may be white, yellow, red, brown, green, gray, blue, or transparent and colorless. Precious opal, which is not found in Kansas, shows a beautiful play of colors and is highly prized as a gem stone. Opal cannot be scratched by a knife but is slightly softer than quartz. It is found as a lining or filling in cavities in some rocks, as a deposit formed by hot springs, and as the petrifying material in much fossil wood.

A common Kansas mineral, opal is widespread in the Ogallala Formation in Clark, Ellis, Logan, Ness, and Rawlins counties. This Ogallala opal is colorless to white or gray and is found with a white, cherty, calcareous rock. Some of it is called moss opal because it contains an impurity, manganese oxide, that forms dark, branching deposits like small mosses in the opal (see Plate 28). Moss opal (or moss agate) has been found in Trego, Gove, and Wallace counties. Opalized fossil bones and shells of diatoms are also found in the Ogallala Formation, as is a green opal that acts as a cement in hard, resistant sandstones.

Feldspar (Hardness 6).—The term feldspar applies to all members of a group of minerals composed of aluminum silicates carrying principally potassium, sodium, or calcium. The feldspars are light in color (pink, green, white, and gray), have a glassy or satiny luster, and have good cleavage in two directions, almost at right angles to each other. They cannot be scratched by a knife. Feldspars commonly occur in igneous rocks; granite boulders, which are found in Woodson County intrusives (mica-peridotite), are largely made up of white feldspar and quartz. The feldspar found in sedimentary rocks in Kansas was brought in as pebbles in streams and in boulders of igneous rocks in the glacial till or boulder clay. Feldspar pebbles may be distinguished from quartz pebbles by their good cleavage.

Sedimentary Structures

In addition to rocks and minerals, Kansas has a number of other formations that are best labeled sedimentary structures. These formations, although composed of Kansas rocks and minerals, require additional explanation. Some, such as concretions or cone-in-cone, may be mistaken for fossils. Others, such as geodes, can be spectacularly beautiful. And some, such as ripple marks, give clues about the climate and geology during geologic history.

Concretions

Concretions are masses of inorganic sedimentary materials in other sediments. They are generally harder than the rocks surrounding them, and therefore many of them weather out of the rocks. Concretions may be formed from any of a number of minerals. In Kansas they consist of calcite, limonite, barite, pyrite, or silica, the last in the form of opal, chert,

Figure 44—Sandstone concretions are not only found in central Kansas. This round sandstone formation is located in central Pottawatomie County.

Figure 45—An unusual sandstone formation, in Pottawatomie County.

Figure 46—This septarian concretion from north-central Kansas is about the size of a volleyball.

chalcedony, or quartz. The shapes vary from round to oval, to long and narrow. Many of them have irregular shapes that can be described as lumpy or globular. The smallest concretions are oölites, which are less than two millimeters in diameter (or smaller than the head of a pin); the largest are many feet across.

Concretions are formed at the time the sediment is deposited, shortly after deposition, or after the sediment has hardened. Water, containing chemical elements, deposits the material either in cavities in the rock or around the rock particles, cementing them together to form a hardened mass.

Kansas has many interesting concretions. In the volcanic ash deposits—such as those near Calvert in Norton County and south of Quinter in Gove County—there are concretions of ash cemented with calcite. Loess deposits also have small areas of calcite forming irregularly shaped nodules. Sandstones in the Dakota and Kiowa formations of the central part of the state have many large concretions, which are also the result of calcite cement. The largest of these (at Rock City in Ottawa County) have diameters of up to twenty-seven feet.

The dark shales of the Cretaceous have a special type called septaria, or septarian concretions (see Figure 46). These are large concretions cut by many veins filled with yellow to brown calcite and occasionally other minerals, such as barite, gypsum, sphalerite, or quartz. They are thought to have been formed by shrinkage of concretions, which caused cracking in the outer layers, thus opening spaces in which minerals were deposited. Small septarian concretions are sometimes called thunder-eggs. They can be found in many places; one of the best localities is a half mile south of Hobbie Lake in Osborne County.

Geodes

Geodes, a type of concretion, are crystal-lined cavities in rocks. They are formed by ground water, which deposits minerals in solution on the walls of the rock cavities. This type of deposition usually forms good crystals, most of which point toward the center of the cavity. Geodelike forms occur in fossil shells, and the entire shell may become lined with crystals. The minerals deposited may be quartz, calcite, barite, pyrite, galena, sphalerite, celestite, or dolomite.

Kansas geodes consist mostly of quartz, chalcedony, and calcite. They have been reported near the town of Rock, along the Walnut River in Cowley County, from the rocks on the hill just north of the Walnut River north of the town of Douglass in Butler County, and from Riley, Cherokee, Marshall, Logan, Trego, Brown, and Wallace counties. Many good quartz geodes have been reported from Chase County. Geodes are quite common and can be found in many localities.

Cone-in-Cone

Cone-in-cone is a peculiar structure consisting of nests of cones, one inside another, standing vertically and arranged either in thin beds or at the edges of large concretions. Some cones are less than an inch in height, and others are as much as ten inches high. They have a ribbed or scaly appearance. Most cone-in-cone is composed of impure calcium carbonate, but occasionally the structure has been found in gypsum, siderite, and hard coal.

In Kansas cone-in-cone is abundant in Kiowa shale of Cretaceous age, where it occurs as beds extending laterally for many feet. When eroded or weathered out, it breaks into small pieces that may easily be mistaken for "chopwood," or petrified wood. Although cone-in-cone may look like fossilized wood, it is an inorganic structure, not the result of any fossil. Good specimens may be found in the banks of the Smoky Hill River in Ellsworth County. Cone-in-cone has also been reported from the limestones in Montgomery, Lyon, McPherson, Washington, and other counties.

Mud Cracks and Rain Prints

When muddy sediments are formed in shallow water, they are often exposed at low tides or in dry seasons long enough to permit drying and cracking. Fossil mud cracks are very similar to present-day mud cracks, except that further deposition has filled in the cracks and preserved them. Fossil mud cracks, when uncovered by erosion, may look like a honeycomb of ridges on a bedding plane. Raindrop impressions may be preserved under the same conditions when they make pits in soft sediments.

Casts of Salt Crystals

When salty mud dries, its surface becomes more or less covered with crystals of halite. Many of these crystals are cubes, but some have hollow faces and are known as hopper crystals. As they are covered up by more sediments the salt itself may be dissolved, but the crystal outlines are commonly preserved (filled with mud or silt) and are known as salt casts. These features are found in many shales and siltstones, and they are particularly common in some of the Permian red beds of south-central Kansas.

Ripple Marks

Ripple marks in many sandstones and siltstones are troughs and ridges that look like the ripples in loose sand in a stream, shallow lake, sea, or sand dune (see Figure 47). Because of this similarity, observers can make shrewd guesses about the rock's origin and direction of prevailing winds. Other fossil ripple marks may be similar to the ripples in dune sand, indicating that the sandstone in which they are found was deposited by the wind.

Some of the water-type ripple marks are symmetrical from crest to crest, and some are not. The symmetrical ripple marks have been formed by waves in standing water, whereas the asymmetrical ripples were formed by water currents. Water-current ripple marks will tell observers what direction the water came from, because the gentle slope faces the current and the steep slope is away from it, or downstream.

Tables for Identification of Kansas Minerals

Two tables showing characteristic properties should be useful in identifying Kansas minerals (see accompanying pages). So far as minerals from other states are concerned, the tables may be of no value and actually can be misleading. It is therefore strongly recommended that published tables (such as those in *Dana's Manual of Mineralogy*) be consulted for identification of any non-Kansas mineral.

Before attempting to identify minerals, the beginning collector should have on hand a small bottle of dilute hydrochloric (muriatic) acid (obtainable in any drugstore); a piece of unglazed porcelain, such as a small tile, for streak tests; an inexpensive steel pocketknife; a piece of ordinary window glass; and a chunk of quartz. A candle and a small pocket magnifier are also useful.

Figure 47—Ripple marks in sandstone, Chautauqua County.

Table I. Minerals with Metallic or Submetallic Luster

A. Will mark paper (hardness less than $2\frac{1}{2}$)

Streak	Color	Hardness	Remarks	Name, composition
Black	Black	1–2	May be in radiating fibrous masses.	Pyrolusite, MnO_2
Gray-black	Lead-gray to blue-black	$2\frac{1}{2}$	In cubic crystals with perfect cleavage. May be massive granular. Small globules of metallic lead collect on surface of fragment held in candle flame.	Galena, PbS
Yellow-brown	Yellow-brown (to dark brown or black)	1+	Earthy. Usually much harder. Apparently noncrystalline.	Limonite, $FeO(OH) \cdot nH_2O + Fe_2O_3 \cdot nH_2O$

B. Will not readily mark paper, but can be scratched by knife (hardness $2\frac{1}{2}$–$5\frac{1}{2}$)

Streak	Color	Hardness	Remarks	Name, composition
Black	Brass-yellow	$3\frac{1}{2}$–4	Commonly massive. Associated with dolomite, galena, and sphalerite.	Chalcopyrite, $CuFeS_2$
Black or brownish-black	Black	5–6	Massive, may occur as coatings. Associated with pyrolusite.	Psilomelane, primarily MnO_2
Light brown to dark brown (lighter than specimen)	Brown to black	$3\frac{1}{2}$–4	Perfect cleavage in six directions. Resinous luster.	Sphalerite, ZnS
Yellow-brown	Dark brown to black	5–$5\frac{1}{2}$	Glassy luster. Seemingly noncrystalline.	Limonite, $FeO(OH) \cdot nH_2O + Fe_2O_3 \cdot nH_2O$

C. Cannot be scratched by knife (hardness greater than $5\frac{1}{2}$)

Streak	Color	Hardness	Remarks	Name, composition
Black	Pale brass-yellow	6–$6\frac{1}{2}$	Massive granular. Commonly in striated cubes or pyritohedrons.	Pyrite, FeS_2
	Very pale yellow	6–$6\frac{1}{2}$	Commonly in cockscombs or radiating fibrous structures.	Marcasite, FeS_2
	Black	6	Strongly magnetic. Crystals are small octahedrons.	Magnetite, Fe_3O_4
Dark brown to black	Black	$5\frac{1}{2}$–6	Commonly massive granular.	Ilmenite, $FeTiO_3$
	Black	5–6	Massive, may occur as coatings. Associated with pyrolusite.	Psilomelane, primarily MnO_2
Yellow-brown	Dark brown to black	5–$5\frac{1}{2}$	Glassy luster. Seemingly noncrystalline.	Limonite, $FeO(OH) \cdot nH_2O + Fe_2O_3 \cdot nH_2O$

Table II. Minerals with Nonmetallic Luster

A. Colored Streak

Streak	Color	Hardness	Remarks	Name, composition
Red-brown	Dark reddish-brown to steel gray to black	$5\frac{1}{2}$–$6\frac{1}{2}$	Massive; radiating. Some varieties softer. Coloring matter in some sandstones (brownish-red).	Hematite, Fe_2O_3
Yellow-brown	Yellow-brown to black	5–$5\frac{1}{2}$	Earthy to hard, with glassy luster. Seemingly noncrystalline.	Limonite, $FeO(OH) \cdot nH_2O + Fe_2O_3 \cdot nH_2O$
Light brown	Light brown to dark brown	$3\frac{1}{2}$–4	Perfect cleavage in six directions. Resinous luster.	Sphalerite, ZnS
Pale yellow	Pale yellow	$1\frac{1}{2}$–$2\frac{1}{2}$	Granular, earthy, crystallized. Burns with blue flame, giving sulfur dioxide odor.	Sulfur, S
Light green	Bright green	$3\frac{1}{2}$–4	Radiating, fibrous. Occurs as small specks in some dolomite beds.	Malachite, $Cu_2CO_3(OH)_2$

B. Colorless streak
1. Can be scratched by fingernail (hardness less than $2\frac{1}{2}$)

Cleavage, fracture	Color	Hardness	Remarks	Name, composition
Perfect cleavage in one direction (the micas)	Golden yellow-brown; brownish-red	$1\frac{1}{2}$	As small scales or books. Expands when heated.	Vermiculite, $(Mg,Fe)_3(Se,Al,Fe)_4O_{10}(OH)_2 \cdot 4H_2O$
	Greenish-white; yellowish; colorless	2–$2\frac{1}{2}$	As small scales or books.	Muscovite mica, $KAl_2Si_3O_{10}(OH)_2$
	Dark brown, green to black	$2\frac{1}{2}$–3	As small scales or books.	Biotite mica, $K(Mg,Fe)_3AlSi_3O_{10}(OH)_2$
	Yellowish-brown	$2\frac{1}{2}$–3	As small scales or books, with copperlike reflection from cleavage faces.	Phlogopite mica, $KMg_3AlS_{13}O_{10}(OH)_{10}$
Perfect cleavage in one direction; good cleavage in two directions	Colorless, white, gray, pink	2	In flat crystals, broad cleavage flakes (selenite) or compact massive without cleavage, or fibrous with silky luster (satin spar).	Gypsum, $CaSO_4 \cdot 2H_2O$
Uneven fracture	Pale yellow	$1\frac{1}{2}$–$2\frac{1}{2}$	Granular, earthy, crystallized. Burns with blue flame, giving sulfur dioxide odor.	Sulfur, S
One perfect cleavage, rarely seen	White, reddish, or yellowish	2	Long needlelike crystals; on mine walls, tri-state district.	Goslarite, $ZnSO_4 \cdot 7H_2O$

2. Cannot be scratched by fingernail but can be scratched by knife (hardness 2½–5½)

Cleavage, fracture	Color	Hardness	Remarks	Name, composition
Perfect cleavage in three directions at right angles	Colorless, white, red, blue	2½	Common salt, soluble in water. Salty taste. Granular cleavable masses or cubic crystals.	Halite, NaCl
Cleavage in three directions at right angles (no cleavage if massive)	Colorless, white, bluish-gray, red	3–3½	Crystals rare. Commonly in massive fine aggregates (not showing cleavage) associated with gypsum; massive variety can be distinguished only by chemical tests.	Anhydrite, $CaSO_4$
Perfect cleavage in three directions not at right angles	Colorless, white, and various tints	3	Effervesces in cold acid. Many crystal forms. Chief mineral in limestone. Fibrous, banded, and granular varieties do not show cleavage.	Calcite, $CaCO_3$
Perfect cleavage in two directions at right angles. Imperfect cleavage in third direction	White, blue, yellow, pink	3–3½	Commonly in aggregates of tabular crystals. Heavier than most non-metallic minerals (differentiated from celestite). In sand-barite rosettes.	Barite, $BaSO_4$
	White, blue, red	3–3½	Similar to barite. Distinguished by crimson flame test.	Celestite, $SrSO_4$
Cleavage not prominent	Colorless or white	3–3½	Small splinter fusible in candle flame, producing lead globules. Hard, brilliant luster. Granular masses and platy crystals, associated with galena.	Cerussite, $PbCO_3$

3. Cannot be scratched by penny but can be scratched by knife (hardness 3–5½)

Cleavage, fracture	Color	Hardness	Remarks	Name, composition
One cleavage direction, indistinct	Colorless, white, various tints	3½–4	Effervesces in cold acid, falls to powder in candle flame. May be in radiating, needlelike crystals.	Aragonite, $CaCO_3$
Three perfect cleavage directions not at right angles (rhombohedral	Colorless, white, various tints	3	Effervesces in cold acid. Many crystal forms. Chief mineral in limestone. Fibrous, banded, and granular varieties do not show cleavage.	Calcite, $CaCO_3$
	Colorless, white, pink	3½–4	Commonly in curved rhombohedral crystals with pearly luster. In granular masses as dolomite limestones. Powdered mineral effervesces mildly in cold acid.	Dolomite, $CaMg(CO_3)_2$

	Light brown to dark brown	$3\frac{1}{2}$–4	In cleavable masses or small curved rhombohedral crystals. Also fine granular (without cleavage). Becomes magnetic after heating in candle flame. Occurs in clay iron-stones.	Siderite, $FeCO_3$
Perfect cleavage in three directions at right angles (or no cleavage if massive)	Colorless, white, bluish-gray, red	3–$3\frac{1}{2}$	Crystals rare. Commonly in massive fine aggregates (not showing cleavage) associated with gypsum; massive variety can be distinguished only by chemical tests.	Anhydrite, $CaSO_4$
Perfect cleavage in two directions at right angles. Imperfect cleavage in third direction	White, blue, yellow, pink	3–$3\frac{1}{2}$	Commonly in aggregates of tabular crystals. Heavier than most non-metallic minerals (differentiated from celestite). In sand-barite rosettes.	Barite, $BaSO_4$
	White, blue, red	3–$3\frac{1}{2}$	Similar to barite. Distinguished by lighter weight, crimson strontium flame test.	Celestite, $SrSO_4$
Perfect cleavage in six directions	Yellow, brown	$3\frac{1}{2}$–4	Resinous luster. In small four-sided crystals or in cleavable masses. May be massive.	Sphalerite, ZnS
Cleavage in two directions, rarely seen	White, pale green, blue	$4\frac{1}{2}$–5	Radiating crystal groups and globular forms.	Hemimorphite, $Zn_4Si_2O_7$ $(OH)_2 \cdot H_2O$
Conchoidal fracture	Colorless, white, yellow, red, brown, green, gray, blue	5–6	Seemingly noncrystalline. Hardness less than fine-grained quartz.	Opal, $SiO_2 \cdot nH_2O$
Cleavage rarely seen	Brown, green, blue, pink, white	5	In rounded globular forms or honeycomb masses. Rare rhomb-shaped crystals. Effervesces in cold acid.	Smithsonite, $ZnCO_3$

4. Cannot be scratched by knife but can be scratched by quartz (hardness $5\frac{1}{2}$–7)

Cleavage, fracture	Color	Hardness	Remarks	Name, composition
Two cleavage directions at nearly 90° angles	White, gray, bluish, pink, green	6	In cleavable masses or irregular grains in rocks. Common in stream gravel.	Feldspar, $KAlSi_3O_8$, or $NaAlSi_3O_8$ to $CaAl_2Si_2O_8$

Conchoidal fracture	Colorless white, yellow, red, brown, green, gray, blue	5–6	Seemingly noncrystalline. Hardness less than fine-grained quartz.	Opal, $SiO_2 \cdot nH_2O$
	Gray, light brown, cream, yellow, red, green	7	Waxy to dull luster. May be banded or lining cavities. Cryptocrystalline quartz.	Chalcedony, SiO_2
	Colorless, white, amethyst, variously tinted	7	Crystals are six-sided prisms capped by pyramids. Often massive, coarsely crystalline. Glassy to greasy luster.	Quartz, SiO_2

5. Cannot be scratched by quartz (hardness greater than 7)

Cleavage, fracture	Color	Hardness	Remarks	Name, composition
Conchoidal fracture	Colorless, white, amethyst, variously tinted	7	Crystals are six-sided prisms capped by pyramids. Often massive, coarsely crystalline. Glassy to greasy luster.	Quartz, SiO_2
Uneven to subconchoidal fracture	Brown, red	$6\frac{1}{2}$–$7\frac{1}{2}$	Crystals have many faces of about equal size.	Garnet, Silicates of Al, Ca, Mg, Fe, Mn, or Cr

4

Fossils

by Debra K. Bennett

From the first discovery of buried bones and shells, the human imagination has been stirred by fossils. They are tangible evidence of plants and animals long vanished from our environment, providing a glimpse of creatures that none of us—even our remote ancestors—have ever seen. Without fossils for evidence, who could dream of duck-billed dinosaurs, or others covered with bizarre plates and spikes? Or sharks in the middle of the continent? Or flying reptiles with a wingspan of twenty feet?

Describing the life of these ancient animals and plants is the work of paleontologists, scientists who spend their lives studying and reconstructing the appearance, movement, and habitat of fossilized organisms. In the Great Plains of North America, paleontologists have been active for more than a hundred years. Imagine, if you will, one paleontologist pursuing his search in the High Plains of northern Kansas.

Vignettes of Earlier Fossil Hunters

Wave upon wave, the hot prairie wind ripples the heads of golden buffalo grass ripened by the August sun. The year is 1882. A flatbed wagon creaks across the prairie in Phillips County, just south of the Nebraska border. The driver is Kansas fossil hunter Charles F. Sternberg. Sleepy, he lets the tired horse go where it will. For another mile, it plods eastward. Then the wind shifts, bringing a delightful smell: water. The exhausted animal pulls the wagon over a last hill, then splashes its nose in the shallows of a small creek near a bank. Dull from the heat, Sternberg steps down from the wagon, thinking of dinner and bedroll. But then, blinking not from heat but from surprise, he is met with a sight to gladden the paleontologist's heart: thousands of fossil rhinoceros bones eroding from the bank!

Figure 48—Layered sedimentary rocks exposed in badlands are good places to look for fossils; these are in Gove County. (Photo courtesy J. D. Stewart.)

The year is 1884, and the prairie wind picks up a fistful of grit and hurls it in the face of John Bell Hatcher. Crouched over another newly exposed fossil rhinoceros bone, Hatcher ignores the wind and dirt. The team and scraper, employed for the past month to help remove several thousand cubic feet of sand covering the bones, has gone home for the day. The quarry has been carefully gridded with stakes and string. There is less grit in the air as Hatcher squats in a five-foot by five-foot square, sketching the exact position of another leg bone of the fossil rhinoceros *Teleoceras*. A page in his journal is devoted to each of more than fifty squares. Another fifty squares have not been mapped, but Hatcher is running out of paper. No paper can be bought in Phillipsburg. The supply train will not be in for another month, and Hatcher is to return East before then. He must save enough sheets for labeling crated bones, and for monthly reports to his employer, Professor O. C. Marsh of Yale University. Hatcher's thin knapsack of supplies, containing pencils, quill pens, blotters, ink, and the remainder of the precious paper, lies now not far away.

The gusty wind picks up again, this time carrying not only sand, but paper as well. Hatcher ignores the wind . . . then with a cry, leaps after the paper that blows away. The first detailed map of a Kansas fossil mammal quarry will this autumn day be left incomplete because of a lack of paper.

In the year 1884, a portly, dark-suited professor sits back in an oaken armchair and contemplates with satisfaction his enormous collection of fossil rhinoceros bones. A month ago, the railway company had finished laying the special side track leading to the lower entrance of the newly completed Yale Peabody Museum. Then two boxcars had been rolled up and braked. Museum workers had labored diligently to unload the crates packed that fall in Kansas by Hatcher. The collection lay now largely uncrated. The fossil bones of the rhinoceros *Teleoceras* were interesting, of course, for they showed its short legs and hippolike body. "A lake-dweller, certainly," the professor says happily to himself, "in perfect accord with the theories of the European geologists who claim Kansas was covered by a huge lake when these bones were deposited." Marsh looks over the crates containing bones of more than two hundred animals, ranging in age from newborn to aged. But the professor's real interest lies elsewhere, in a smaller section of the collection: the bones of fossil horses. Teeth and bones indicate some three-toed forms. They fit in nicely with the professor's conception of evolution as a linear succession of forms, predirected toward a culmination in the modern horse. "Another missing link, another intermediate form," he muses as he sits at his writing desk. The succession of fossil horses had become more complete each year, the structural changes from ancestral to descendant forms better documented with each study. Marsh sighs as he contemplates publishing the last word on horse evolution. Then he picks up a pen and dips, intending to write a letter of termination to Hatcher. The professor will be sending Sternberg to collect dinosaurs in New Mexico next year, but first a few questions for the young collector, his former student at

Figure 49—Paleontologists O. W. Bonner, L. D. Martin, and J. D. Stewart, of the University of Kansas Museum of Natural History, excavate a giant Cretaceous clamshell containing rare fishes. (Photo courtesy K. N. Whetstone.)

Yale, about the rocks in and near the rhinoceros quarry. What had Hatcher said about their looking like river and ancient soil deposits, rather than like lake sediments? Preposterous! And yet, Hatcher had seemed so meticulous . . . and what *would* all those three-toed horses be doing swimming far out in a lake?

The sun is setting over the lip of the quarry. The wind dies down, and the mournful lowing of a lone, skinny steer drifts down to the rhinoceros bones exposed below. Sand shifts and slides down the bank face at the head of the creek, which now forms a steep quarry backwall. The year is 1936, and George F. Sternberg, hands and forearms covered with plaster of Paris, looks into the sunset. This has been another dry summer, and the grass is thin in the pasture above. "Not much to hold the earth together," Sternberg observes as he mixes another batch of plaster to protect the bones about to be crated. His dad had warned him about slumping in the overhanging lip of the backwall. But the daily reports and forms to fill out for his employer, Childs Frick of the American Museum in New York, and the dry weather, good for plaster-jacketing bones, put his mind off the task. Then, too, the landowner had been away looking for factory work and hadn't brought over the Ford and scraper. Sand shifts as Sternberg pats on another handful of plaster. Suddenly the shift becomes a slide, and with a rumble and a choking cloud of dust, the overhang comes down, covering three weeks of quarrying.

The prairie wind still blows, and the afternoon sun glares down on two graduate students dressed in worn blue jeans and army surplus field shirts. The year is 1979. The students crest a hill, and then, struggling in the wind with a large map containing not quite the necessary detail, proceed to argue about the location of the old Sternberg rhinoceros quarry. One gestures toward the head of a draw fronting the south side of the hill. "The archives at Yale and in the Frick lab at the American Museum say it's down there. And so does that old map of Hatcher's. Maybe we'll find the old quarry dumps." Sliding over the bank lip somewhat east of their course up and out of the wash, they send gray sand swishing to the dry creek bed below. "The creek banks are lumpy—too lumpy for undisturbed ground," she thinks. A crunching snap greets their landing in the soft sand below. "Hey, I *found* it," she says.

Once again the rhinoceros quarry will be worked for its fossil treasures.

Answering the Main Question

For every fossil quarry in Kansas, historical vignettes like these could be written. For fossils do not dig themselves out of the ground. Times change, scientific theories become more testable, tools improve; paleontologists are not the same, do not have the same training, as the Sternbergs, Hatchers, and Marshes of the 1880s.

Paleontology today requires training in both biology and geology. Most paleontologists study anatomy, physiology, embryology, taxonomy and systematics, evolutionary theory, sedimentary geology, stratigraphy, and field mapping. Many have computer language or graphics skills. Most also have an interest in philosophy or the history of science. But one thing has not changed over a hundred years of North American paleontology: the main question. Whether paleontologists study fossil clams and find a practical outlet for their expertise in employment for an oil company, or whether they study for the sheer joy of discovery the locomotory capabilities of short-legged rhinoceroses, the main question remains the same: what was life like in the past?

This chapter is designed to answer "the main question." The answer was discovered through scientific research over the past hundred years in Kansas and made clearer through the efforts of hundreds of laboratory and

Figure 50—Fossils and models at the Sternberg Memorial Museum, Fort Hays State University.

field workers. It is both scientifically useful and tantalizingly incomplete.

Paleontologists can discover, describe, illustrate, and analyze only fossils that are, in fact, found. A trip through any Kansas paleontological museum leaves many people wondering why so many specimens are needed, but the diverse specimens in the exhibits are only a small portion of the research collection. Unfortunately the research collections, large as they are, represent only a tiny fraction of all Kansas fossils. But what may be most surprising is that, of all the animals and plants that ever lived in Kansas, only a tiny fraction, perhaps one-tenth of one percent, have actually been fossilized. Several biological factors conspire to limit the number of organisms that become fossilized. Because soil minerals have an affinity for cellulose, bone, and shell, an organism possessing hard parts is much more likely to be fossilized than one without them. Because burial is required for the chemical processes that produce fossilization, an organism living in a river delta, marsh, or swamp is more likely to be fossilized than one living in

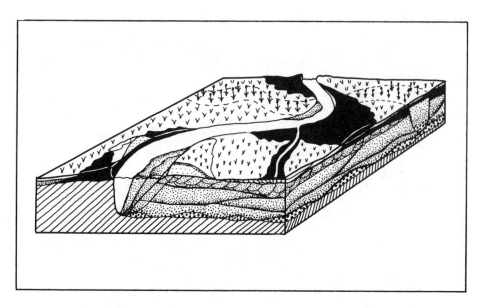

Figure 51—The history of a winding river, or of any natural feature, is recorded in sediments. The structure of sediment layers can tell paleontologists what kind of environment existed at the time of burial of fossil animals and plants.

sand dune country or in the freshet of a mountain stream. Finally, to become fossilized the remains must not be scavenged and eaten by any of the abundant forms that Nature provides for her self-renewing cleanup. Thus, paleontologists constantly seek more fossils, for they know that the earth's past can best be reconstructed with the help of collections that are as complete as possible.

Following an order apparent in nature, paleontologists are divided into three groups. The first group, paleobotanists, studies the remains of fossil plants, including plant pollen, plant microfossils (spores and seeds), and plant macrofossils (whole cones, trunks, chunks of wood, bark, and leaves). The second group studies animals without backbones. These invertebrate paleontologists seek both the rare fossils of soft-bodied invertebrates as well as their much more common shelled relatives. The third group, vertebrate paleontologists, studies animals with backbones. These include remains of fishes, amphibians, birds, reptiles, and mammals.

Over the past hundred years, an increase in both theoretical and practical knowledge has helped the "search efficiency" and interpretive power of Kansas paleontologists. For example, we now know that Hatcher was right about the origin of the sediments containing Tertiary rhinoceros bones: they were mainly river and not lake deposits. Paleontologists have verified Hatcher's observation by examining the sedimentary patterns in the ancient sands, and comparing these patterns to those produced in modern lakes and rivers. This knowledge produces much predictive power. For example, a fossil found in a long, sinuous river deposit can help predict where other fossils in the same river deposit may lie (Figure 51).

Marsh's concept of evolution, however, is unlikely. Marsh believed that ancestral forms change in a predirected way until they achieve an apogee in a living descendant. In the case of the horse, this kind of evolution is now known not to occur, for there are thousands of extinct horse species, many of which were more efficient runners or grinders of grass than the living horse. We now know, too, that because of the sheer abundance of existing but uncollected fossils, and because of a need for more paleontological research, the "last word" about any fossil form is unlikely to be written for many years.

Figure 52—Fossils are not often preserved neatly. This drawing shows a crushed and twisted skull of the Cretaceous fish *Cimolichthys*. Careful checks of its anatomy may someday permit paleontologists to reconstruct its appearance in life.

On the practical level, the pioneering efforts of Charles Sternberg and his sons have greatly improved fossil-collecting technique for field workers all over the world. Paleontologists can now reduce almost to zero the chance of shattering one of these rare treasures between quarry and laboratory. Once in the laboratory, it is now possible to recover bones from even the most intractable clay matrices. To Kansas paleontologist Claude W. Hibbard goes the credit for inventing the technique of wet-screening (sifting clay sediments under water) for recovering small fossils. Small vertebrates, such as mice and shrews, are very sensitive to small climatic changes, and require specific temperature and moisture levels for their survival. Thus the presence of these tiny fossils gives a precise picture of the climate near a fossil site at the time of death of the mice and shrews it contains. This is why paleontologists search for old rhinoceros quarry dumps. Even after the large rhino bones were removed, the surrounding sand and earth, heaped up by early excavators, still contains a wealth of information about life in the distant past.

Maps are also now available, showing both the geological relationships and the topographic shape of any part of the Kansas countryside. Finding fossils is not a random procedure. Vertebrate paleontologists may decide to search a certain portion of the Cretaceous outcrop in western Kansas, and because they have a map, they can find and systematically search it. Invertebrate paleontologists can decide to look for articulate brachiopods, extinct shelled organisms useful in oil exploration, in the Oread Limestone; because they have maps, they can avoid searching where others have already documented the presence of these organisms. But how do paleontologists know which rock units to search for which fossils? Can dinosaur bones be expected in the Oread Limestone? The answer to this question is no and involves both biological and geological knowledge.

Over the past hundred years, understanding of the succession of Kansas rock layers, or stratigraphy, has greatly improved. Figure 1 shows the Kansas stratigraphic column, which, read from bottom to top, shows rock layers deposited between 2.5 billion years ago and the Quaternary period. Wherever they occur, these rock layers always have the same mutual relationships, and the position of any one layer in the stratigraphic column can be determined by observing adjacent layers. Each rock layer is distinctive in both chemical composition and in the selection of fossils typically included within it. By noting these three factors—relative position, chemical composition, and included fossils—field workers can determine the relative age of any layer in the column. Unless affected by later deformation, the older rock layers, deposited first, always lie lower than younger ones.

Since about 1950 and the development of peacetime uses of radioactive technology, it has been possible to determine the absolute age in years of rock layers or of fossils included in these rocks. The technique compares the amount of "undecayed" or radioactive substance remaining in a sample with an amount of undecayed substance estimated originally to have been present. Different elements, such as carbon, argon, or uranium, which commonly occur in nature in their radioactive form, are used for this comparison. Because the half-lives, or decay rates, of different elements are of different lengths, different segments of the stratigraphic column can be dated by each. By reading a geologic map (Plate 32), fossil hunters know where different rock layers are exposed on the surface of the state. Rocks belonging to each time span are color-coded on geologic maps.

Results of paleontological surveys of each rock layer, representing corresponding time intervals, have now been compiled from all over the world. These surveys demonstrate an important fact: in the past, just as now, organisms were adapted to specific environments. Adaptations show up in the fossil record as differences in the sizes and shapes of the hard parts of animals and plants, but the soft body parts of the living creatures had adaptations, too. For example, sharks, with their soft cartilaginous skeletons, flipperlike limbs, and "swim-pump" breathing apparatus, do not live in forests. Nor do bears, with their heavy fur coat, small terminal nostrils, relatively small lungs, and long clawed feet, ply the ocean depths. These are extreme examples, but many species have very narrow tolerances for heat, light, moisture, minerals, organic nutrients, and living space. If all of these environmental factors, and more, do not exist within an organism's tolerances in a given place, it will not survive in that locale.

The past century of paleontological research has shown that each species occurs only in certain rock layers (and thus exists only for a finite period of time) and that each species is adapted to a specific environment. As a result, dinosaur fossils cannot be expected in the Oread Limestone. This rock layer is about 280 million years old, yet the known range of rocks containing dinosaur bones is from 240 to 63 million years. Thus the Oread Limestone was deposited before dinosaurs existed on the earth. Furthermore, the animal species included within the Oread Limestone—as well as the chemical composition and the depositional pattern of the rock itself—tell Kansas paleontologists that it was deposited in a sea basin. But no dinosaur yet discovered possesses the paddle-shaped limbs and other adaptations characteristic of marine vertebrates. Dinosaurs can thus be expected only in Kansas rocks of the right time span and the right environment. Rocks of the appropriate time span do exist in Kansas, but of these Mesozoic rocks, most were deposited by a sea that covered part of Kansas at that time. Dinosaur bones are therefore rare in Kansas, but marine vertebrates of the Mesozoic—such as mosasaurs and plesiosaurs—are common here.

A Trip Through Kansas Prehistory

One way to understand fossils is by taking a tour up the Kansas stratigraphic column (Figure 1). Reading from bottom to top, our trip will emphasize changes in land forms and life forms that are known to have occurred over about the last 800 million years. Except for a small area of southeastern Kansas, only rocks of Middle Pennsylvanian age (about 300 million years old) and younger are exposed on the surface in Kansas. Our knowledge of older rocks comes from drill hole samples and from surveys of rocks of equivalent age that are exposed on the surface in other parts of the world.

Deep within the earth, several thousand feet below the rolling prairie, lie the oldest rocks yet studied in Kansas. These rocks, 600 million years old and more, contain few fossils. These Precambrian rocks began either as the spew of volcanoes and hot magma bodies, or as sandy and muddy sediments. The latter may have once contained fossils, but time has seen much change in these sediments. None is now either chemically or physically the same as in its beginning, and few fossils survived these changes. All these rocks were squeezed and partially melted and bent to form mountains, then eroded and deposited and squeezed and melted again to form mountains from the roots of mountains. The rock cycle—uplift and erosion, with corresponding rise and fall of nearby sea levels—continues today.

The inevitable erosion of the lofty summits of mountains continued across the arbitrary time-boundary we now call the beginning of the Cambrian. All the time from the beginning of the Cambrian to the present, 600 million years, constitutes less than one-seventh of the whole history of the earth. These younger rocks have been folded into mountains and washed into the sea fewer times than those of the Precambrian. The fossil record contained within them is therefore much more complete than that of the Precambrian. And there is another reason why paleontologists know more about life after the Precambrian: suddenly, for reasons still not well understood, animals with shells and other hard parts became common, and these have been preserved abundantly in the rocks.

A complete stack of all rock layers ever deposited is not present at any single place in the world. A complete stratigraphic column can be compiled only by careful comparison of shorter partial stacks from all over the world. Thus it is reasonable to expect breaks in the stratigraphic record preserved in Kansas. Rocks of lower and middle Cambrian age are missing not only in Kansas but over all of the midcontinent of North America. But even absence contains information: this gap represents a long span of time in which the rate of erosion exceeded the rate of deposition of new sediment. Paleontologists thus infer that dry land was exposed in Kansas at this time.

Rocks of late Cambrian age do occur in Kansas. They were deposited on an uneven, hilly surface formed by erosion during the middle Cambrian. The chemical composition and texture of these middle Cambrian rocks suggest an upland origin. No fossils are found within these rocks, for land-dwelling life probably did not yet exist on the earth. Soon, however, the exposed land was submerged, and at the close of the Cambrian period a thick series of marine limestones was deposited.

The Cambrian fossil record contains remains of every major group of

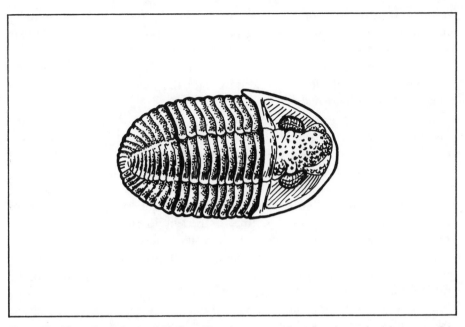

Figure 53—The arthropod group includes spiders, insects, scorpions, horeshoe crabs, lobsters, crayfish, and trilobites. Whole trilobites, such as the *Phillipsia,* are rare finds from the Kansas Pennsylvanian. (After Moore, Lalicker, and Fisher, 1952.)

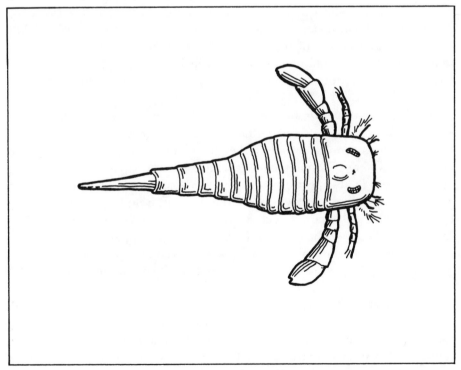

Figure 54—Eurypterids grew to over six feet in length during the Silurian. (After Moore, Lalicker, and Fisher, 1952.)

animals, with one exception: vertebrates. The shallow seas were inhabited instead by jellyfishes (of which, surprisingly, there are a number of fossils preserved as body impressions), sponges, various types of worms (whose fossil record consists mostly of trails and burrows), brachiopods, bryozoans, and primitive sea-lily forms. The dominant life form was the trilobite, a relative of today's horseshoe crab.

Marine deposition continued into the Ordovician period. During early Ordovician time, a variable group of beach sands and sandy, offshore limestones was deposited in Kansas. These sediments indicate that Kansas may have been near the shore of a shallow sea lying to the south. This sea saw the first of the scorpionlike eurypterids (see Figure 54). Crinoids (sea-lilies) and straight-shelled cephalopods (relatives of today's curve-shelled *Nautilus*) became abundant. Colonial or reef-building corals first became common. Sponges reached a peak of diversity, and echinoids (sand dollars and their relatives) appeared as simple, globular, stemmed forms. Scummy aggregations of algae floated in shallow nearshore waters, soaking up sunlight. Swimming among this garden of undersea life came the first chordates—jawless, unarmored, and limbless. They are the ancestors of the oldest fishlike vertebrates.

Yet this was an unstable period of earthquakes and volcanic explosions. Rock layers deposited atop the Kansas Ordovician beach sediments show eroded surfaces, indicating that slowly pulsating, mountain-building activity caused the land to rise and fall several times. Between early and middle Ordovician time, some of the early Ordovician and Cambrian sediments were folded into mountains, down from which came new deposits of beach sand. The surrounding seas deposited limy sands and fine limy muds. The waters were deep.

In them, eurypterids became giants up to six feet long. Clams were common. Brachiopods of certain families clung together in reeflike formations, building upward to track the life-giving sunlit zone of the ever-deepening waters. Globular coral colonies abounded, and shallow-water algaes formed food-capturing, sticky mounds composed of matlike layers. Sea snails (gastropods) crawled over the reefs, grinding limy deposits with their rasplike mouth parts to get at the juicy algae or corals hidden within. Straight- and curved-shelled nautiloids were the strongest swimmers of the time and searched the midwaters for prey. Trilobites were still common. Jawless vertebrates now became armor-plated, perhaps to defend themselves from attack by the eurypterids.

In the nearshore shallows, seaweeds—rootless plants without true stems or leaves, yet still more complex than the colonial algaes—washed in the waves. Some washed ashore, and a few survived there. Late in the Silurian, some developed a vascular system of stiff internal tubes that could hold the short, fleshy plant upright and could carry water to its upper parts. Dry land had at last been invaded.

Fossilized plant spores are present in late Silurian and early Devonian

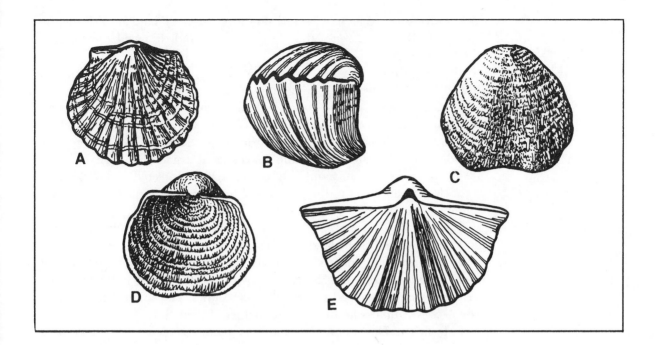

Figure 55—Brachiopods have the upper shell smaller than the lower shell (as in B). Compare brachiopod and mollusk shell symmetry (see Figure 71). A, *Meekella,* top view. B. *Meekella,* oblique side view. C, *Juressania,* top view. D, *Juressania,* bottom view. E, *Neospirifer,* top view. (After Moore, Lalicker, and Fisher, 1952.)

Figure 56—Vertebrates possess cartilage or bone in their skeletons. Here an armored Devonian *Dinichtys* chases the shark *Cladoselache.* Despite its fearsome aspect, *Dinichtys* lacked true teeth. The shark belongs to a much more advanced group.

rocks of Kansas. Valleys were eroded in newly uplifted land at this time, but by middle Devonian time, the sea again rose and drowned the formerly upland valleys.

In these drowned valleys lived an abundance of sunlight-loving reef dwellers. Tall, flowerlike crinoids flourished in diverse communities also containing blastoids; horn, tube, and chain corals; brachiopods; sponges; gastropods; trilobites and nautiloids. Yet other developments of the Devonian are most important for the world as we know it. Vascular plants developed roots, true stems, and leaves, and could therefore grow far inland, away from a constantly high water table. Horsetails, some growing as much as twenty feet tall, club mosses, and ferns were the first groups to fully conquer the land environment.

What walked through the early Devonian fern forests? No amphibian had yet moved across the land, but arthropods (animals with chitinous outer skeletons, including spiders, scorpions, and cockroaches) developed land-breathing adaptations and came ashore with the plants. Still in the sea, some vertebrates now acquired paired limbs and jaws. Devonian fishes became the fastest swimmers and fiercest predators of their day.

Late in the Devonian, the land was again invaded. Certain kinds of lungfish, of a family that gulped air when the water became stale, crawled on belly and bony flippers out onto the land. For the first time, a vertebrate suffered under the full force of gravity. Unsupported by water, its muscles strained to gulp air while its skin dried in the fresh breeze. The first amphibian snapped up an insect and struggled back into the water. Like the first land plants, many of the first land vertebrates did not survive in this harsh new environment.

Despite these Devonian land conquests, marine life was still far more abundant. Sharks plied the waters, and the heavily armored bony fish became stronger swimmers. In Kansas and over most of the western half of North America, the Devonian began a long interval of marine deposition that was followed in the Mississippian period by deposition of a thick series of shallow-water limestones filled with the remains of reef life. There is little evidence to indicate that the land surface emerged in Kansas until late Mississippian time, when erosional surfaces can again be found. Sandstones and shales deposited in river deltas and marshes became more frequent while marine limestones and shales become less frequent at this time. The land emerged gradually, the sea making many brief inundations.

Rocks of Pennsylvanian age in Kansas are distinguished by cyclothemic sequences—regularly repeating sequences of limestone-shale-sandstone layers indicating brief rises and falls of sea waters and the land's repeated upward pulsations. Before deposition of Pennsylvanian rocks in Kansas, the eroding surface of Mississippian and older rocks was deformed by renewed pulsating earth movements. These determined the rate at which the sea receded and dictated the deposition of the cyclothems.

In the Mississippian and Pennsylvanian seas, shelled life forms reached

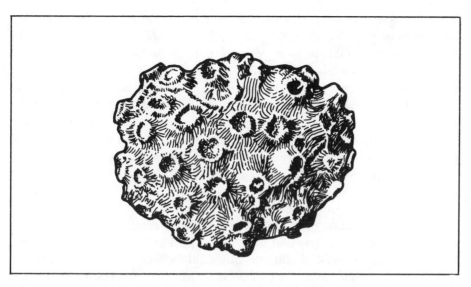

Figure 57—*Pachyphyllum,* a colonial coral. This specimen is about eight inches across. (After Moore, Lalicker, and Fisher, 1952.)

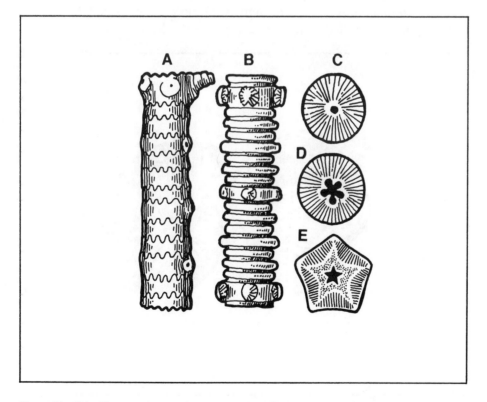

Figure 58—Crinoid stem parts are the most common Kansas Pennsylvanian fossils. A, stem with smooth, interlocking rings. B, stem with rough, noninterlocking rings. C,D,E, end views of rings with different shapes. Rings vary from one-eighth inch to about one inch in diameter. (After Moore, Lalicker, and Fisher, 1952.)

an astounding diversity. Many Carboniferous limestones of eastern Kansas are excellent sites for collecting these invertebrates, which can often be obtained in beautiful condition. Very abundant—in fact, sometimes largely composing a given rock unit—are the large, shelled protozoans called Foraminifera. These look like grains of wheat and are the remains of surface-floating microorganisms. Horn-of-plenty–shaped solitary corals as well as colonial rock and chain corals are all common. The latter two types built great reefs that provided a fastening ground for many of the shelled forms. The screwlike, sticklike, and sheetlike bryozoans are known by the small lipped pores found on one surface of the skeleton only. Each small pore once contained a single bryozoan animal that waved tiny tentacles in water currents to capture microorganisms for food and retreated into its "apartment" in case of danger. Crinoid and echinoid parts are very common in these rocks (crinoid stem parts look like little wheels or donuts). Echinoids are relatives of starfish and sand dollars. Their skeletal parts sparkle when broken, for each is formed of a single crystal of calcium carbonate. Mollusks, especially gastropods (snail relatives), cephalopods (octopus relatives), and pelecypods (clam relatives) are moderately abundant. Blastoids, free-floating relatives of the rooted crinoids, and trilobites are less common but present. Through the shallow-water reefs of the Kansas Pennsylvanian patrolled primitive Xenacanth sharks.

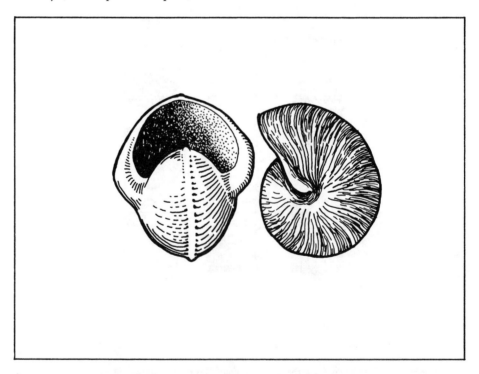

Figure 59—A particularly beautiful Pennsylvanian snail, *Bellerophon*. Its shell is planispiral and it superficially resembles an ammonoid (Figure 69), but most ammonoids are much larger than the one-half inch wide *Bellerophon*.

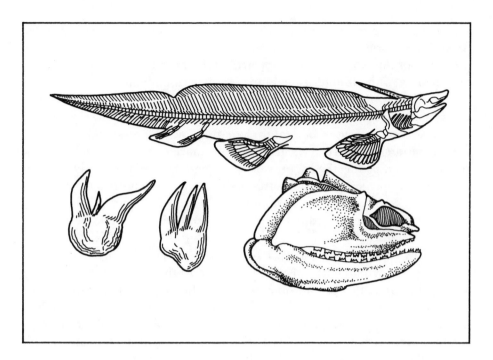

Figure 60—The body outline, skeleton, teeth, and skull of a primitive shark, *Xenacanthus*. The head, spine, and teeth of this form are found in Kansas Pennsylvanian rocks, but other parts are rarely fossilized.

The greatest abundance of shelled forms in these sediments is, however, among the brachiopods. Brachiopods differ from pelecypods in body symmetry despite the fact that both have two shells. Today pelecypods are dominant, but during the Kansas Pennsylvanian, brachiopods were represented by hundreds of genera classified in at least six orders. Each order contains forms different in shell shape, shell construction, and inferred shape of the gills of the living animal (the gills often left marks on the inside of the shell).

Yet during the Pennsylvanian period, terrestrial sediments slowly began to predominate. In Pennsylvanian rocks, the first strong evidence of differences in the environment from region to region within Kansas can be detected. In eastern Kansas, sediments were deposited in separate valleys between which rose ridges and hill tops, but deposits in southern Kansas became thick enough to cover even these divides. West and north, Pennsylvanian sediments thin and change to marine chemistries, indicating the presence of a sea.

Great forests developed on land during part of the Mississippian and Pennsylvanian times. These are now preserved as coal seams. In them, club mosses grew to over sixty feet high, while horsetails lined the wetlands. The palm-shaped *Sigillaria* and scaly-trunked *Lepidodendron* filled vast tracts, while huge *Cordaites* and tree ferns shaded Kansas valleys. Horsetails (*Calamites*) grew tens of feet tall.

Living and rotting plants provided food in the forests for land-dwelling cockroaches, dragonflies, flies, crayfish, and snails, which in turn were preyed upon by spiders and scorpions, while these in turn were eaten by descendants of the flipper-legged lungfish: the amphibians.

For the first time, life of the fresh- and brackish-water rivers, marshes, lakes, and deltas left a fossil record in Kansas. Fresh- and brackish-water sharks swam upriver. Coelacanth lobe-fins, not closely related to the amphibian ancestors, thrived in brackish estuaries. Mound-building algal colonies encamped at the mouths of rivers and captured not only food on their sticky upper surfaces, but fortunately for paleontologists, also the bony remains of dead amphibians washing downstream. These algal mounds have been collected and dipped in acid to recover the amphibian bones encased within.

Late in the Pennsylvanian, a brackish-water estuary in southeastern Kansas was surrounded by teeming life. Along the shores of this estuary lived a new kind of vertebrate—long-necked, sharp-toothed, and more agile than the amphibians living nearby: a reptile. When upstream pools feeding the estuary dried up, amphibians living there died out, for their naked eggs, laid in water, dried and the embryos died within. But the leather-shelled and large-yolked eggs of *Petrolacosaurus,* laid on land, hatched to produce another generation of insect-loving predators. Little *Petrolacosaurus,* named for the Rock Lake Shale from which its bones come, is the world's oldest diapsid reptile, belonging to the same group as the great dinosaurs.

Plate 29 is a reconstruction of a quarry containing plants and animals from the Pennsylvanian. The composition of carbon sediment enclosing fossils from this quarry in the Garnett area shows that lightning often started fires in the pine forests covering the uplands. The waters were a mixture of fresh water moving downriver to the sea and salt water moving up with the tides. In this estuary, marine animals such as cup corals (left foreground of the color plate), clams (pink siphon tubes in right foreground), and bryozoan colonies (gridded structure in the right foreground) could live much as their relatives today live in mangrove swamps. The coelacanth fish *Synaptotylus* probably preferred the saltier bottom water. Pine branches and cones *(Lebachia),* ginkgo fronds (*Dichophyllum,* leaning on bryozoan), and cordaite branches (under coelacanth) fell to the muddy bottom to become fossilized among twisted roots of seed ferns (right) and tree ferns (left). The armored amphibians depositing their yellowish egg mass in the upper, fresh waters are dissorophids, ancestors of modern frogs. A bright green *Petrolacosaurus* snaps up a dragonfly from a perch atop a *Lepidodendron* log. A *Lepidodendron* cone floats in the foreground. A cockroach and scorpion crawl over the seed fern trunk while a spider spins among tree fern fronds. The dark green, bananalike leaves of another tree fern form a backdrop for the light green of a giant horsetail rush. Relatives of these coal-forest giants live today only in the tropics.

Alternating layers of limestone and siliceous rock (chert or flint) form the

Flint Hills of Kansas and contain abundant remains of shelled marine forms from the Permian period. West of the Flint Hills, red beds overlying these flinty limestones form the Red Hills geographic province. Little life seems to have survived in the dry salt plains of the Kansas Permian. Sail spines or jaws of *Dimetrodon* and other fin-backed reptiles are occasionally found in these deposits. In brackish pools surrounded by salt-loving plants, certain lungfish dug burrows to wrap themselves in mud and body mucus to wait out the dry seasons. Paleontologists have collected the skeletons of some that never emerged to another rainy season. Sharks of advanced types swam upstream. In the sea, the mechanics of the biting jaws of bony fishes had improved, and the fish had shed some of the heavy weight of armoring scales. Hundreds of species of these faster-swimming, quicker-turning, snap-biting fish evolved.

Exposures of rocks of Mesozoic age are confined to the western two-thirds of Kansas, but rocks of Triassic and Jurassic age are either absent or very rare in the state. Erosional processes predominated during the whole of the earlier Mesozoic in Kansas, and the land surface was exposed.

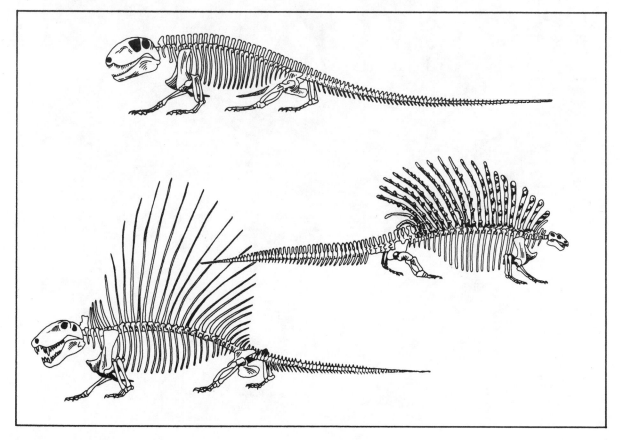

Figure 61—Pelycosaurs, such as *Sphenacodon* (top), *Edaphosaurus* (middle), and *Dimetrodon* (bottom), are known from the Kansas Permian. Their skull structure reveals that they are more closely related to mammals than to dinosaurs.

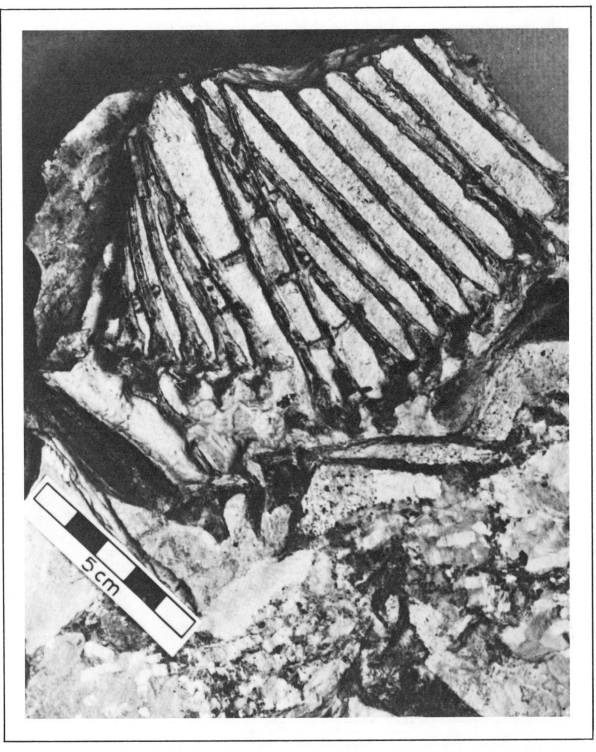

5cm

Figure 62—Pelycosaur spines, as they appear after careful paleontological preparation has removed covering rock layers. (Photo courtesty Robert Reisz.)

With the beginning of the Cretaceous period, the sea rose relative to the land. An arm of the sea stretched northward from the Gulf of Mexico and gradually became connected to an arm of the sea reaching southward from Alaska. The continent along this line gradually sank, and thus the sea basin deepened across western Kansas. The oldest Cretaceous rocks in Kansas are upland or shoreline deposits laid down in a marine backwater or river delta environment. Marine muds and sands were deposited over these land-generated rocks as the sea basin deepened and the water spread wider. The Niobrara and other chalk formations of the Upper Cretaceous Series were formed by the accumulation of billions of tiny shells of one-celled microorganisms, which over millions of years lived, died, and drifted to the bottom

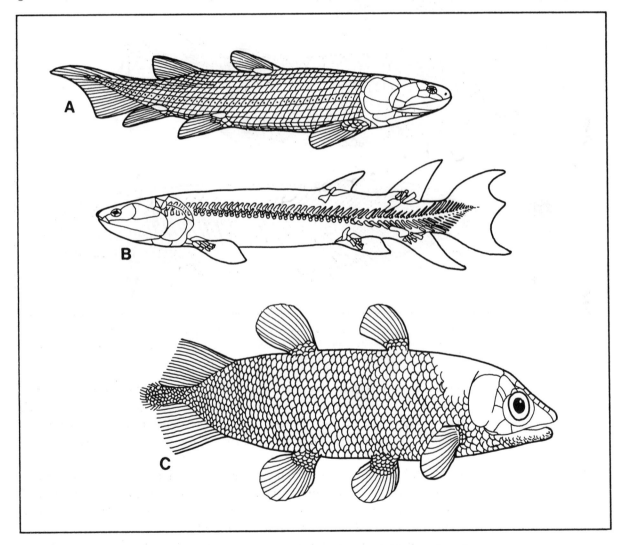

Figure 63—Lobe-finned fish. A, crossopterygian, *Osteolepis,* of the Devonian, showing heavy scales and fleshy-based fins. B, skeleton and body outline of a crossopterygian, *Eusthenopteron,* of the Devonian, showing the bony fin structure. From this group are derived the land-dwelling vertebrate groups. C,

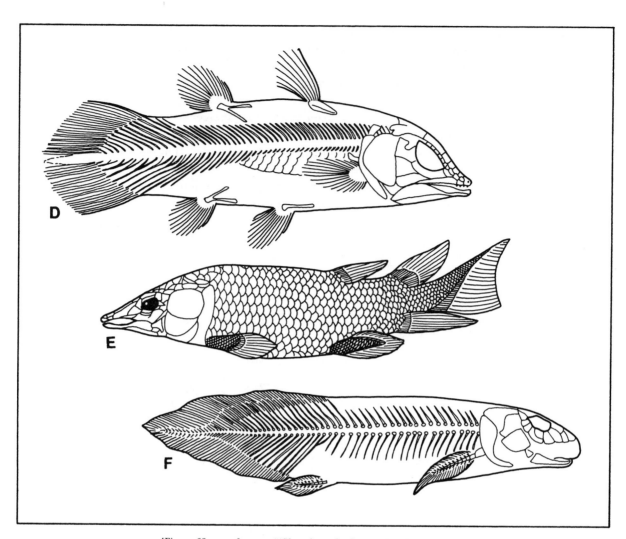

[Figure 63 cont. from p. 117] coelacanth, *Synaptotylus,* Cretaceous, showing scale pattern and fringed fins. D, skeleton of coelacanth *Macropoma,* Pennsylvanian. E, a dipnoan lungfish, *Griphognathus,* with a pointed snout. F, dipnoan lungfish skeleton.

of the Cretaceous sea. The slow rain of these tiny limy shells accumulated to 750 feet thick in parts of northwestern Kansas. For reasons still not well understood, most of the Cretaceous shoreline deposits were subsequently removed by erosion, while the marine Cretaceous of Kansas largely remains.

Much had happened on land during the earlier part of the Mesozoic. Reptiles had diversified and crocodiles, turtles, lizards, and dinosaurs had become dominant. Mammals, still small and secretive, scurried among bushes and the roots of a new kind of plant—the coniferous or "evergreen" tree. High up in the branches, early birds swooped and flapped, snapping up flying and crawling insects in their toothed beaks. They maneuvered with long, feathered tails.

Reptiles evolved from the water-loving amphibians, but many reptiles had by the Cretaceous returned to a fully aquatic life. The dominant predators in the Kansas Cretaceous sea were the mosasaurs, relatives of today's lizards and snakes, and the plesiosaurs, a group now wholly extinct. Pterosaurs, or flying reptiles, soared over the Cretaceous waters. Pterosaurs (also known as pterodactyls), mosasaurs, and plesiosaurs all preyed on an abundance of fishes, for by this time fully modern bony fish, closely related to living forms, had evolved. A diversity of sharks, some huge and active predators with rows of sharp teeth, and some slower bottom-dwelling forms with flat, mollusk-crushing teeth, also competed for food.

Shallows along the shoreline of the Kansas Cretaceous sea seem to have been sunny and quiet. The fine bottom sediment was easily stirred by the smallest current and was so soft that animals moving over it on legs would quickly have sunk in. The only large invertebrates that seem to have been able to avoid suffocating on this soft bottom were huge clams called *Inoceramus,* whose shells were very thin-walled and comparatively light, and which grew up to five feet in diameter. The clams apparently lay with one of their shells flat against the bottom. On their shells grew a variety of barnacles and smaller oysters, forming "minireef" communities. The fossilized skeletons of schools of small, rare fishes (*Leptecodon, Omosoma, Kansius, Caproberyx*) have been found *inside* the clam shells (see the reconstruction of Cretaceous fish in Plate 30). A picture of an unusual reef community emerges. The bright-colored fishes may have gleaned food from the giant clam gills, or may have dodged between the clam shells for shelter from larger predatory fish or reptiles. Certainly, there was little other protective cover available.

The Cretaceous was a time of great diversity for the cephalopods, especially the ammonoids. One ammonoid shell from the South Dakota Cretaceous preserves the bite-marks of a mosasaur that unsuccessfully attempted to eat it. Giant squids, up to six feet long, also plied the waters. These cephalopod forms were carnivores, like almost all other large animals of the time. The ecosystem in the Cretaceous sea was thus a delicate one. Paleontologists know of few plant-eaters in this environment; rather, the ecological structure consisted of plankton, the floating microorganisms that constituted food for some fish, and carnivores, of which there were many. Thus, any climatic change affecting the plankton would have had serious repercussions on all sea life.

On the basis of many lines of evidence, great climatic changes are known to have occurred all over the world at the end of the Mesozoic era. A great erosional surface separates Mesozoic and Cenozoic rocks in Kansas. The Cretaceous sea drained away as the land rose once again. This land rise coincided with a major phase in the formation of the Rocky Mountains.

Rocks of early Cenozoic age do not occur in Kansas. The long period of erosion that began at the end of the Cretaceous continued until Miocene time. Miocene rocks in Kansas belong to the Ogallala Formation, widely exposed along stream banks in the western half of the state. In middle Miocene time, a second phase of uplift in the Rocky Mountains formed the Front Range, and this uplift continues today. The Ogallala Formation consists primarily of deposits that eroded from the rising Front Range and were carried to Kansas by wind and streams.

At one time, the Cenozoic deposits west of the Flint Hills were part of a nearly continuous body of sand and silt, stretching from South Dakota in the north to New Mexico in the south, and from the Front Range to as far east as central Kansas. This body of sediment was wedge-shaped, thicker to the west, and composed of low hills and broad, shallow river valleys. Erosion has subsequently removed much of this sediment, but remnants of this old surface still remain and are known as the High Plains. On the High Plains surface in Meade County, you can walk where extinct Kansas rhinoceroses once walked. Looking down in the present-day valleys, you can guess at the huge volume of Ogallala sediment that has already been washed to the sea.

The Kansas Miocene was populated by many species of mammals. From their small beginnings in the late Triassic, and after extinction of the dinosaurs at the end of the Mesozoic, mammals became larger and more numerous. By the end of Eocene time, representatives of all the orders of mammals were in existence. By Miocene time, Kansas was occupied by some forms now extinct but many other kinds that are still living. The climate was milder and wetter than today. Flowering plants, which had first appeared in the Cretaceous, formed forests of deciduous trees. Late in the Miocene, the forests began to thin, and a new plant, grass, became common. The landscape was savannalike, as in Africa today. Forests lined the watercourses, while on the upland divides, clumps of trees were frequent. Plant pollen, seeds, leaves, diatoms (tiny, glassy-shelled micro-organisms), and—occasionally—volcanic ash, blew in the Kansas wind and collected on pond bottoms.

In and around one diatom-producing pond in Wallace County, the bones of many Miocene animals have been found. The pond was a reedy oxbow, shallow and filled with clear water, connected to a large river nearby by marshy land and a small creek. Let us travel back in time six million years, to look at a Kansas pond, some of whose inhabitants are similar to animals alive today, while others have been extinct for at least five million years (Plate 31 is a reconstruction of one moment in time at the Rhinoceros Hill site).

We may imagine the day beginning. Sunrise over the pool brings the chirp and grumble of bullfrogs and spring peepers. Salamanders slither over fallen logs. A hawk soars overhead, while a heronlike bird dips its bill in the shallows to search for crayfish. Schools of sunfish flicker among the reeds, while minnows look on. At the upstream end of the pool, willows dip their long branches in the water among cattails and mulberry thickets. A small weasel-like mammal slips into the water to try the fishing, while the raccoonlike ringtail scuttles up the branches of a cottonwood to finish eating

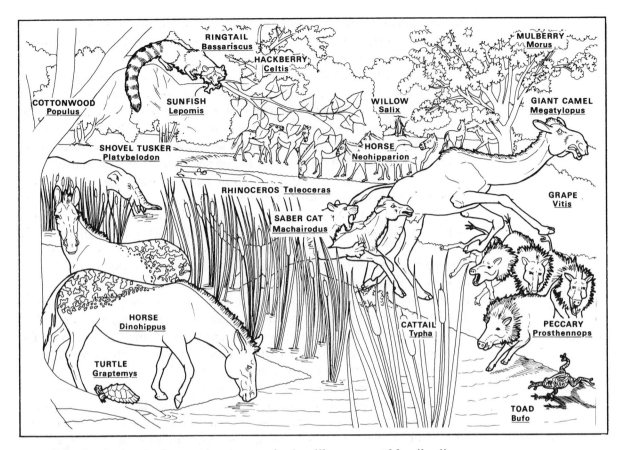

a sunfish. At the head of a nearby draw, a badgerlike mammal busily digs a burrow in the sandy bank. A cougar-sized cat passes downwind. Bending branches and grunting with pleasure, a shovel-tusked proboscidean ambles down to the water to dig among the shallows for roots. A long-necked giant camel turns from its drink to eye its gangly calf. The short-legged rhino *Teleoceras* submerges all but its nostrils in the center of the pool while it cools itself and waits to climb out for its night feeding on nearby leaves and grasses. Farther upland, out past the hackberry and elm forest growing on the lip of the divide, another, longer-legged rhinoceros, *Aphelops,* searches the forest edge for edible leaves. The slender, swift llama *Hemiauchenia* lives here too.

As the sun rises higher, a toad dives into the pool under the watchful gaze of a one-toed horse. Pecarries shove each other down to the water to drink. A turtle is warily balanced on a cottonwood root. Noisily trampling in the shallows, a herd of slender-legged, three-toed horses appears on the far bank.

Suddenly, the mother giant camel bawls an alarm. Her calf has strayed too far, and from the bushes comes a tawny blur: a saber-cat. In seconds, powerful clawed forelegs press back the calf's neck, and the slashing canines do their work. The mother camel flees, stumbling over pecarries. The cat

Fossils
121

drags its kill away. Six million years later, paleontologists in Kansas are able to reconstruct the scene from the fossil evidence preserved in muds from the pond bottom and shoreline.

After the Miocene, the climate of Kansas gradually became much drier. Grass became more widespread, and the savannalike broken forests just described became the open prairies of today. By the beginning of the Pleistocene, Kansas was a land of rippling grass. Deposits of this period, chiefly unconsolidated sand and gravel, are widespread but discontinuous. Then came the great glaciers. The northeastern part of Kansas was overridden by glacial ice at least once during the Pleistocene.

Most people think of the Pleistocene as a time of intense cold, but recent scientific evidence shows that, except near the ice margin, the North American climate was more even than today, with less difference in temperature between the summer and winter months. As ice fronts advanced southward, plants that had formerly grown in the north escaped by seeding southward each year, until at times of maximum glaciation, the Kansas forest contained species that today grow hundreds of miles north. The same applies to animal species of these times; all living things not able to live near the cold ice front were compressed toward the equator. By contrast, at times of glacial retreat, southern species such as armadillos, sloths, and giant tortoises often moved northward.

Many Pleistocene mammals were larger than their closest living relatives. In Kansas are found remains of moose, deer, elk, and bison species larger than those now living. Horse is found in the older part of the Pleistocene. But some Pleistocene mammals are now extinct—*Castoroides,* a giant beaver; *Cervalces,* the stag-moose; *Symbos,* the fronted musk-ox; and the two proboscidean types, the mammoth and the mastodon. The mammoth was a grass-eating form, taller and longer-legged, living in herds on the open prairie, while the mastodon was a forest dweller, seeking out leaves and bark.

The end of the Pleistocene saw an extinction of many of these large mammals, while some small mammals also became extinct. The fauna and flora of today are much reduced from the Pleistocene. Paleontologists do not agree about the causes of extinction, nor why the Pleistocene extinction seems to have been selective. Yet the Pleistocene, many workers believe, is still going on; we are living in it, despite the fact that the present arbitrary time interval is called the Holocene. The climatic fluctuations of today can be understood much more clearly with a Pleistocene perspective. Future research in Kansas paleontology will certainly attack the problem of Pleistocene extinctions.

Some Fossil Finds

Thousands of species of fossil animals and plants are known from Kansas rocks. Here are descriptions of some common kinds that an amateur fossil hunter might discover.

Figure 64—Skull top and antler bases of the Kansas Pleistocene stag-moose *Cervalces*. Distance between antler tips in an unbroken skull would have exceeded five feet. (Photo courtesy Carl D. Frailey.)

Paleozoic Invertebrates

Invertebrate fossils are by far the most common from Kansas. They occur in marine and terrestrial rocks all over the state and range in age from Mississippian to Pleistocene. The invertebrates discussed here are only a sample of the fascinating diversity of these forms. Most of those illustrated here come from the Pennsylvanian of eastern Kansas.

Foraminiferans.—These are unusually large one-celled animals that possess shells. These protozoans were fossilized in huge numbers during the Pennsylvanian. Foraminiferan shells are complex, made up of a sectioned and spirally coiled sheet of calcium carbonate. The shells are football-shaped and often show a rough outer surface, making them look like grains of wheat. They are thought to have lived by floating at the surface of the sea, feeding on even smaller floating microorganisms.

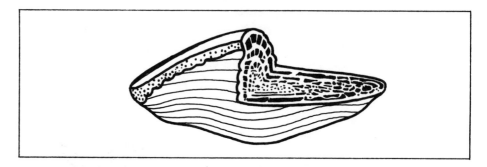

Figure 65—An enlarged and sliced-open view of a shelled, one-celled animal called a fusulinid. This view shows the complex inner shell structure. Billions of these tiny fossils, which look like grains of wheat, can be found in rocks of the Kansas Pennsylvanian and Permian. (After Moore, Lalicker, and Fisher, 1952.)

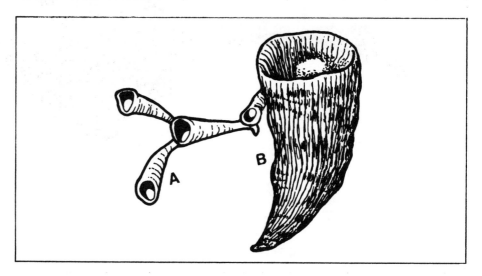

Figure 66—Pennsylvanian corals. A, *Aulopora,* chain coral, one inch long. B, *Lophophyllidium,* solitary coral, one-half to four inches long. (After Moore, Lalicker, and Fisher, 1952.)

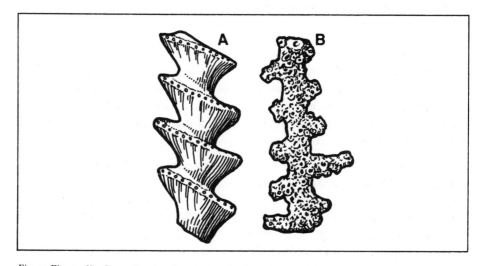

Figure Figure 67—Pennsylvanian bryozoans. A, the spiraling *Archimedes,* one-half inch long. B, a branching bryozoan, about one-half inch long. (After Moore, Lalicker, and Fisher, 1952.)

Corals.—These are solitary or colonial reef-building animals. The small, tentacled, coral animal or polyp secretes an outer coating of calcium carbonate. The polyps feed by extending tentacles out into circulating water. Small animals floating by are numbed by nerve cells in the coral's tentacles. Then the tentacles retract, and the prey is pulled toward the polyp's centrally located mouth.

Corals belong to the same zoological group as jellyfish, sea anemones, and hydras. All possess radial symmetry, meaning that their body plan is structured around a central axis. Coral reefs build upward to track the sunlight zone, for many coral species also derive food from algae living within the polyps' tissues. The algae require sunlight, for they are plants.

Bryozoans.—These are animals that also live colonially and secrete a protective skeleton of calcium carbonate. They do not, however, build large reefs. Instead, they attach their colonies to the reef itself or to shelled animals living on the reef. Bryozoan skeletons possess pores for the individual bryozoan polyps. The pores are located on one side of the skeleton only. This helps paleontologists distinguish them from corals.

Crinoids.—These are flowerlike, stalked animals possessing many frondlike tentacles. Crinoids belong to the same zoological group as starfishes and echinoids (sea cucumbers and sand dollars). All have skeletons made up of thousands of blocks of calcium carbonate. Each block is constructed from a single crystal of the material. When a block is broken, the broken surface sparkles.

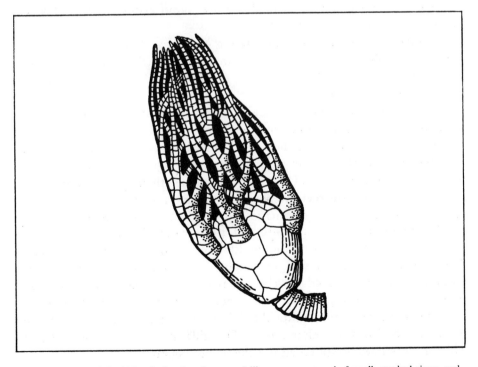

Figure Figure 68—Crinoid head, showing the tentaclelike arms composed of small, stacked rings, and the cuplike body composed of roughly pentagonal plates. (After Moore, Lalicker, and Fisher, 1952.)

Crinoid stem parts, which look like small donuts or wheels (Figure 68), are especially common in Pennsylvanian rocks of Kansas. Crinoid heads look like flowers. The bulbous lower part of the head is made up of hundreds of pentagonal plates. The tentacles are made up of thousands of small donut-shaped rings. Crinoids are occasionally fossilized whole, but since the skeletal parts were held together in life by the soft body of the animal, sea currents usually scattered the crinoid's plates after it died.

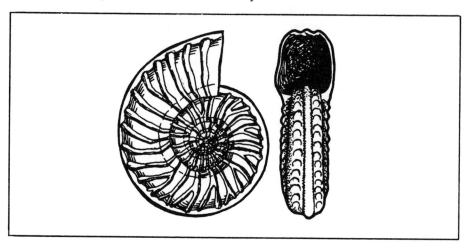

Figure 69—Ammonoid, showing planispiral coiling. Compare to gastropod in Figure 59. Shell diameter one-half inch to one foot. (After Moore, Lalicker, and Fisher, 1952.)

Mollusks.—Several sorts of mollusks are found as fossils in the Kansas Pennsylvanian.

Cephalopods are relatives of the living *Nautilus* and *Octopus*. Ammonoids are extinct cephalopods. They are rare in the Kansas Pennsylvanian. These tentacled forms were strong swimmers, plying the midwaters and surface for fish and smaller cephalopods.

Gastropods include both the sea and land snails. Snail shells are coiled, as are the shells of ammonoids, but the coiling plan is different. Ammonoids are planispiral—that is, all their shell coils lie in a single plane. A few gastropods *(Bellerophon)* are also planispiral (Figure 59), but in most gastropods the first whorl of the shell lies higher than the last and largest whorl. Gastropods are reef-dwellers. They have rasping mouth parts and can grind coral or algal skeletons to capture the algae or polyps within.

Pelecypods include both salt- and fresh-water clams. These animals commonly live on reefs or on the muddy bottoms of seas, rivers, and lakes. They possess a short, fleshy, muscular "foot" that digs down into the bottom and serves to hold the clam in place. Many clams can also move (slowly) from place to place with their foot. Clams live by partially opening their shells and extending a tubelike siphon. They pump sea water through the siphon and past the clam's gills. The gills filter out organic material, while indigestibles are pumped out the siphon again.

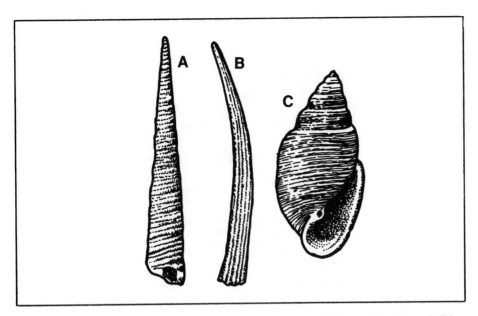

Figure 70—Snails of the Kansas Tertiary. A, *Turritella*. B, *Dentalium*. C, *Actaeon*. (After Moore, Lalicker, and Fisher, 1952.)

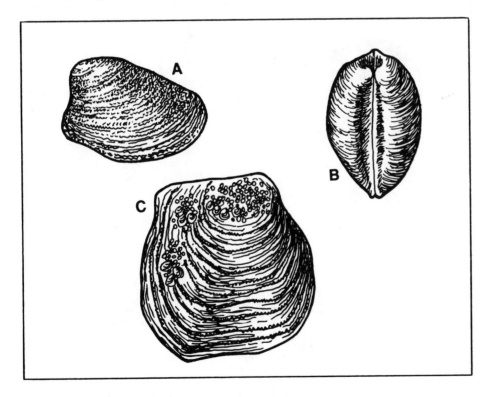

Figure 71—Clams belong to the mollusk group. Note difference in shell symmetry between these and brachiopods (Figures 55 and 73). In clams, both shells are equal sized. A, top view, *Nuculopsis*, Tertiary. B, end view of same. C, *Inoceramus*, top view, a giant, thin-shelled Cretaceous species, reaching four to five feet in diameter. (A and B after Moore, Lalicker, and Fisher, 1952.)

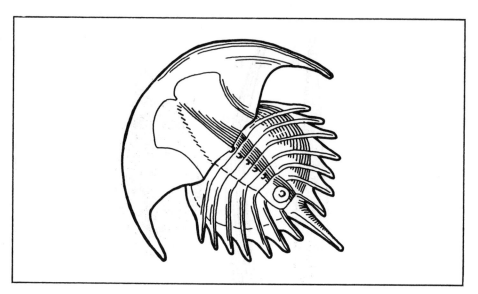

Figure 72—This fossil horseshoe crab, *Prestwichianella,* from the Pennsylvanian, resembles a trilobite but belongs to a separate group (see Figure 53).

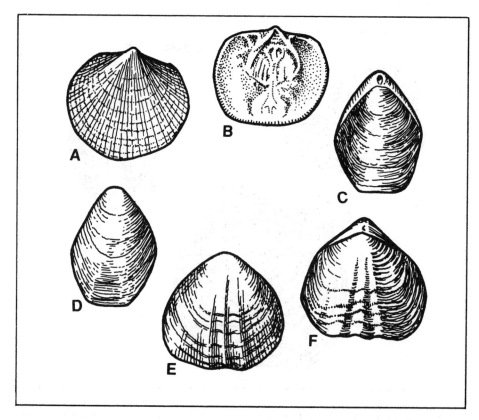

Figure 73—Brachiopods. A, *Rhipidomella,* top view. B, *Rhipidomella,* view of inner surface of upper shell showing impressions made by the gills. C, *Dielasma,* bottom view. D, *Dielasma,* top view. E, *Nudirostra,* top view. F, *Nudirostra,* bottom view. (After Moore, Lalicker, and fisher, 1952.)

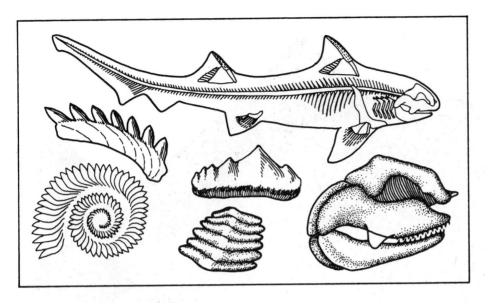

Figure 74—The body outline, skeleton, teeth, and skull of hybodont and hybodontlike sharks, which have sharp or spiraling teeth near the front of the jaws but crushing plates near the back of the jaws.

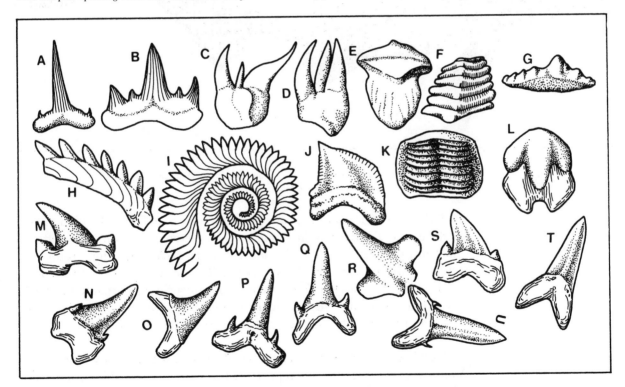

Figure 75—Common Kansas shark teeth. A,B, *Cladodus* (Pennsylvanian). C,D, *Xenacanthus* (Pennsylvanian). E, *Petalodus* (Pennsylvanian). F,G, *Orodus* (Pennsylvanian). H, *Edestus* (Pennsylvanian). I, *Helicopryon* (Pennsylvanian). J, *Squalicorax* (Cretaceous). K, *Ptychodus* (Cretaceous). L, *Rhinobatos* (Cretaceous). M, *Cretolamna* (Cretaceous). N, *Otodus* (Pennsylvanian). O, *Cretoxyrhina* (Cretaceous). P, *Scapanorhynchus* (Cretaceous). Q, *Leptostyrax* (Cretaceous). R, *Squatirhina* (Cretaceous). S, *Otodus* (Pennsylvanian). T, *Cretoxyrhina* (Cretaceous). U, *Isurus* (Cretaceous).

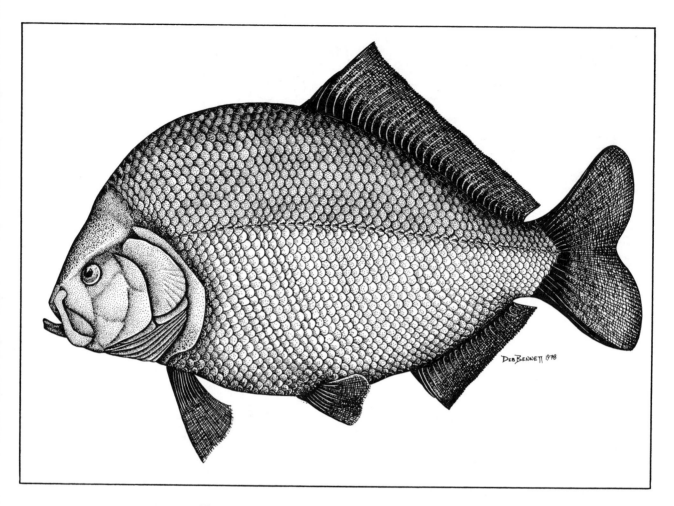

Figure 76—An advanced teleost fish from the Greenhorn Limestone of Nebraska (Cretaceous). (Reproduced by permission of John D. Chorn.)

Trilobites.—These animals are rare in the Kansas Pennsylvanian. They are the extinct relatives of the living horseshoe crab and belong to the group called arthropods, animals with jointed outer skeletons. This group also includes scorpions, spiders, and insects. Trilobites were bottom-dwellers that burrowed through soft sediment while they filtered organic material. The trilobite's distinctive three-part armor permitted fairly flexible movement and protection. Trilobites did not survive into the Mesozoic.

Brachiopods.—These animals reached astounding diversity during the Pennsylvanian, and are well represented in Kansas rocks. Six orders were common. Brachiopods, like clams, have two shells, but differ in body symmetry. Like clams, brachiopods also filter organic debris from the water, but they lack a siphon. The brachiopod is not mobile as an adult and may be fixed to soft substrate by a long, weak "foot," by projecting shell spines, or by cementing deposits.

Sharks, Rays, and Chimaeras.—The skeletons of these animals contain no bony material, but instead are made of the same soft flexible cartilage that holds out the end of your nose and keeps your ears from sagging. Because they are soft, shark skeletons are rarely fossilized. But the fossil record of sharks in Kansas is good, because both shark teeth and shark skin denticles (denticles are toothlike structures embedded in shark skin) are made of hard bone and enamel. Their bony fin spines are also commonly found.

Paleozoic sharks of Kansas include torpedo-shaped, meat-eating forms with spines projecting vertically from near the back of the head, and sharply pointed teeth (Figure 75). These sharks are often found in suspected fresh- or brackish-water deposits. Cretaceous rocks yield the sharp teeth of meat-eating sharks and the flattened, knoblike tooth batteries of mollusk-crushing sharks. While predaceous sharks have torpedo-shaped bodies to aid them in fast swimming, the mollusk-eaters often had flattened bodies to aid them in patrolling the sea bottom.

Teleosts.—These are advanced bony fish (Figure 77). Their remains are found in Cretaceous and Tertiary deposits of Kansas, and can be collected either as slabs containing whole fish skeletons or by screening for individual bones.

Teleosts possess true paired fins that have a unique construction. A short bone with a ball joint attaches the mobile fin to the fish's body. Most of the fin is supported by long, thin, parallel, jointed rays. There are two pairs of fins: a pectoral ("arm") pair and a pelvic ("leg") pair. The paired fins of teleosts are located relatively close to the head. The teleost tail fin is also supported by thin, jointed rays.

The jaw structure of teleosts is their most advanced feature. Jaws of primitive fishes could open and shut like a scissors. These simple jaws were lined with sharp teeth. But the front part of teleost jaws is excluded from the gape, and the jaws are so hinged together that when the teleost mouth is opened, the front part of the jaws shoots out automatically. This creates a strong suction just in front of the jaws. The teleost can capture prey without biting it. Once prey have been sucked into the teleost's mouth, backslanting teeth within prevent the prey from wriggling out.

The body shapes of teleosts are varied. Some, like the Cretaceous *Urenchyles,* are snake-shaped eels. Many are torpedo-shaped predators. Some are deep-bodied and flattened from side to side; these often consume plant material or are reef-biters. All teleosts have lighter-weight scales than primitive bony fish. They are agile and fast swimmers and today are the dominant form of ocean life.

Lobe-fins.—Lungfish fossils are found in Paleozoic sediments of Kansas. Scales and teeth are the most common remains, but burrows containing whole skeletons have been found in Permian pond deposits.

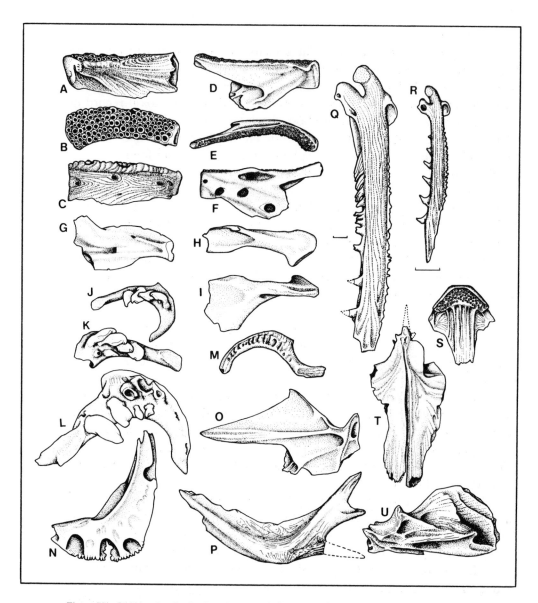

Figure 77—This is what fossil teleost bones look like when they are recovered from Tertiary deposits by wet-screening. A,B,C,P,Q,R,T,U, catfish bones of *Ictalurus*. D,E,F,N,O,S, sunfish bones of *Lepomis*. G,J,K,L, minnow bones of *Semotilus* and *Notropis*. H,I,M, sucker bones of *Catostomus*. None of the bones on this plate are bigger than one-half inch long, except Q, which measures two and one-half inches. J is less than one-eighth inch long.

Kansas lungfish are not closely related to the group that gave rise to the amphibians. Two common Paleozoic genera, *Sagenodus* and *Gnathorhiza*, are related to the living African and Australian lungfish. These forms burrow in the muddy bottoms of ponds and go into a state of torpor to survive dry conditions. A third genus, *Synaptotylus,* is a coelacanth related to the seagoing "living fossil" discovered recently in the Indian Ocean. All Kansas lungfish were fresh-water forms with an omnivorous or carnivorous diet.

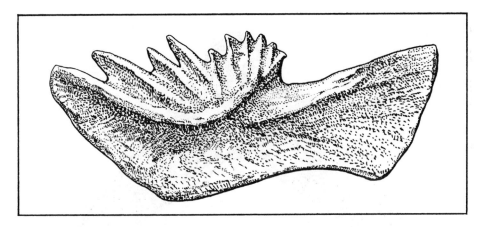

Figure 78—Tooth plate from the lungfish *Sagenodus*. Actual length about two inches. (See lungfish skeletons in Figure 63.)

Figure 79—Frogs, toads, salamanders, and the wormlike caecilians are living members of the amphibian group, but extinct forms are much more varied. Top view of Kansas lower Permian *Acroplous,* an active predator. Specimen about one and one-half inches long. (Photo courtesy Ronn Coldiron.)

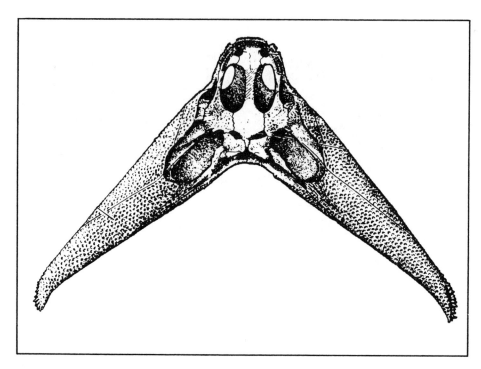

Figure 80—*Diploceraspis* (shown) and *Diplocaulus* are Great Plains Pennsylvanian and Permian amphibians possessing hornlike skull projections. Skulls can measure a foot between horn tips.

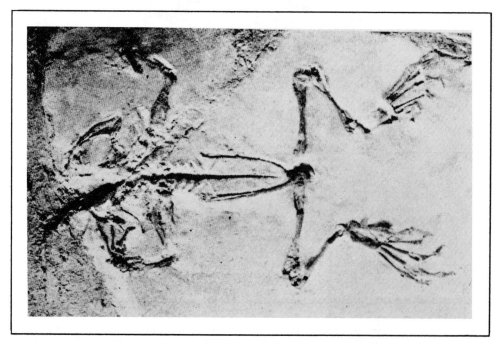

Figure 81—Remains of the Tertiary toad *Scaphiopus* as preserved in a diatomite slab. This genus, the spadefoot toad, still lives in Kansas.

Amphibians.—Kansas rocks of Paleozoic and Tertiary age contain amphibian fossils (Figures 79–81). Paleozoic amphibians belong to groups now extinct, while the Tertiary forms are the more familiar frogs and salamanders. Amphibian remains have been recovered from Pennsylvanian shales, Permian red beds and shales, and Pennsylvanian algal mounds. Tertiary pond deposits occasionally produce slabs containing a whole skeleton, but Tertiary remains are usually recovered by screening.

Typical Paleozoic genera include flat-skulled, froglike forms *(Trimerorhachis)*, slender, fairly agile insect-eaters *(Acroplous)*, and flattened pond dwellers. *Diplocaulus* possessed short, weak legs, a long body, and a bizarre skull with two long, backwardly projecting ''horns.'' It probably never left the water.

Tertiary forms include salamanders, frogs, and caecilians. Caecilians are wormlike, burrowing forms; fossils of these legless animals are rare. A selection of typical Tertiary amphibian bones is pictured. All the fossil amphibians of Kansas were pond or pond-edge dwellers.

Turtles.—Kansas has produced a large variety of turtle remains from Cretaceous and Tertiary deposits (Figures 82–84). The Mesozoic forms are seagoing, while those from the Tertiary are fresh-water pond forms or land-dwelling tortoises. Thick, square shell plates are the most common fossil remains in Kansas Tertiary deposits. Whole skeletons or isolated skulls are rare and scientifically very valuable.

Turtles are reptiles belonging to a group distinguished by solid skull structure. Most fossil and all living turtles possess a beaked face but no teeth. Their upper shells are formed from flattened and extended ribs covered with thickened bony plates. Their lower shells are formed from an expanded breastbone and scutes. Turtle plates generally possess grooves on one or both surfaces.

The limbs of land-dwelling and pond turtles have short toes and permit overland travel. Seagoing turtle limbs have elongated toe bones bound together into paddles of flesh. These turtles leave the water only to lay their eggs. Most turtles are omnivorous. The presence of large land tortoises in some Pleistocene fossil localities indicates that freezing probably did not occur at those sites, for the giant tortoises cannot burrow into the ground to escape the cold, and they will die in freezing temperatures.

Mosasaurs.—Mosasaurs were large, seagoing reptiles of the Mesozoic. Their remains are abundant in Cretaceous deposits in Kansas. All skeletal parts, and even skin impressions, have been found.

Mosasaurs are relatives of snakes and lizards. They had elongate bodies and paddle-shaped limbs. They probably never came ashore, not even to lay their eggs. They probably reproduced as many modern snakes do, by hatching eggs within the mother's body and giving birth to relatively highly developed young. Once born, the young could quickly swim to the surface for their first gulp of air.

Mosasaurs' jaws were double-hinged, and their skull construction was flexible and light. Thus, like many snakes, they could consume large prey. They were fast swimmers, and their blunt, conical teeth could crush and hold ammonoids, squids, or fish. Mosasaurs had large eyes and were probably active during the day. They are found in deposits formed far out in the middle of the Kansas Cretaceous seas.

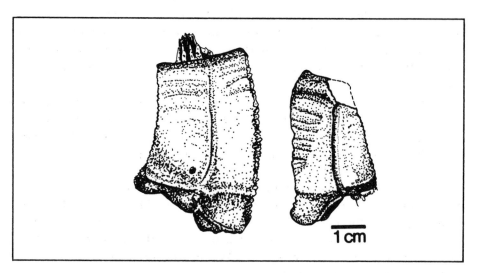

1 cm

Figure 82—The most common Kansas Tertiary fossil—bits of turtle shell. Turtles belong to the reptile group. These plates are about two inches long.

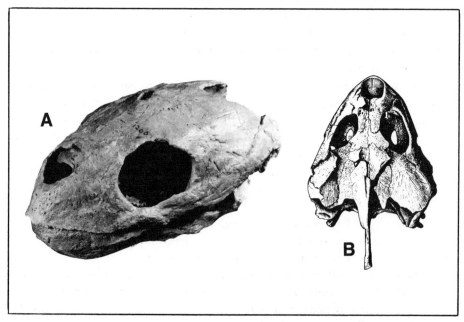

Figure 83—Whole turtle skulls, like these beautiful specimens, are rare. Turtles have beaks but no teeth. A, *Desmatochelys*. B, *Toxochelys*. Both from the Kansas Cretaceous. (Photo courtesy John D. Chorn.)

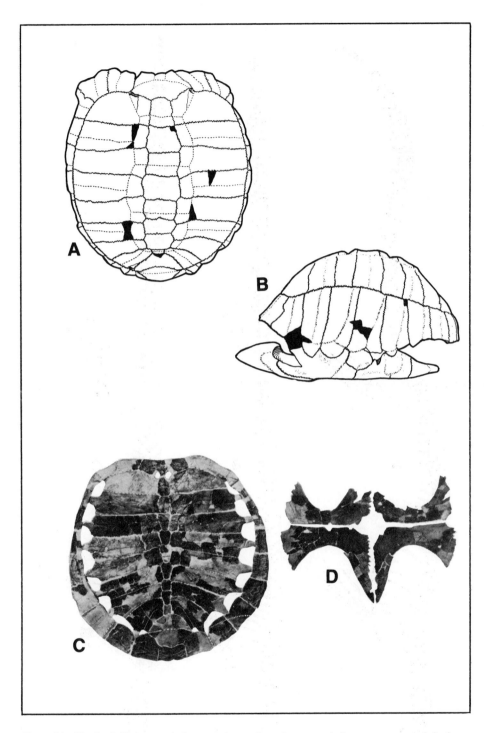

Figure 84—Turtle shell plates articulate together to form the upper shell, or carapace, and the lower shell, or plastron. A,B, *Geochelone,* shell in top and side views; length two to five feet. C,D, *Toxochelys,* shell length up to two feet. C is the carapace, D is the plastron. *Geochelone* is a giant land tortoise of the Kansas Pleistocene, while the relatively lightweight shell of *Toxochelys* suited it for its marine existence in the Cretaceous. (C and D after R. Zangerl, 1953.)

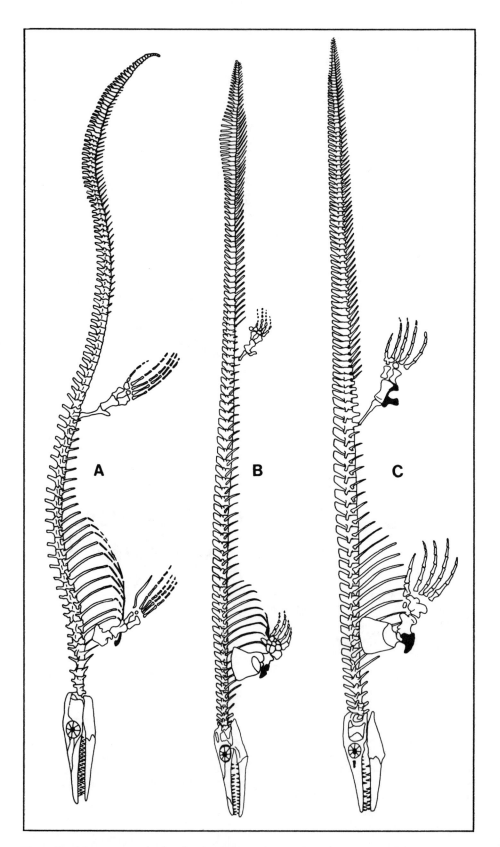

Figure 85—Mosasaurs are closely related to living snakes and lizards. A, *Tylosaurus*. B, *Clidastes*. C, *Platecarpus*. All are abundant in the Kansas Cretaceous.

Figure 86—Mosasaur teeth are distinctive—conical, pointed, and set in sockets.

Plesiosaurs.—Plesiosaurs also lived in the Kansas Cretaceous seas (Figure 87). They came in two varieties: long-necked and short-necked. The long-necked forms had small skulls; the short-necked forms had long skulls.

Plesiosaurs had rowboatlike bodies propelled by large, stiff paddles. A slender tail acted like a rudder, while immensely powerful abdominal muscles arched the body during paddle strokes. The plesiosaur progressed through the water somewhat like a swimmer doing the breaststroke.

The slender neck of the long-necked plesiosaurs was incredibly flexible. "Anchored" at the surface far out at sea, these plesiosaurs could easily swivel their necks backward for a look aft and could extend the head far out from the body to dip prey from among unsuspecting schools of fish. The short jaws of these forms were replete with long, interlocking, daggerlike teeth that could easily have speared several small fish at a single bite.

The short-necked plesiosaurs seem to have had more slender bodies and could probably swim fairly fast. They probably pursued their prey. Their long, snapping jaws were lined with short, needlelike teeth. All plesiosaurs had large eyes and were probably day hunters.

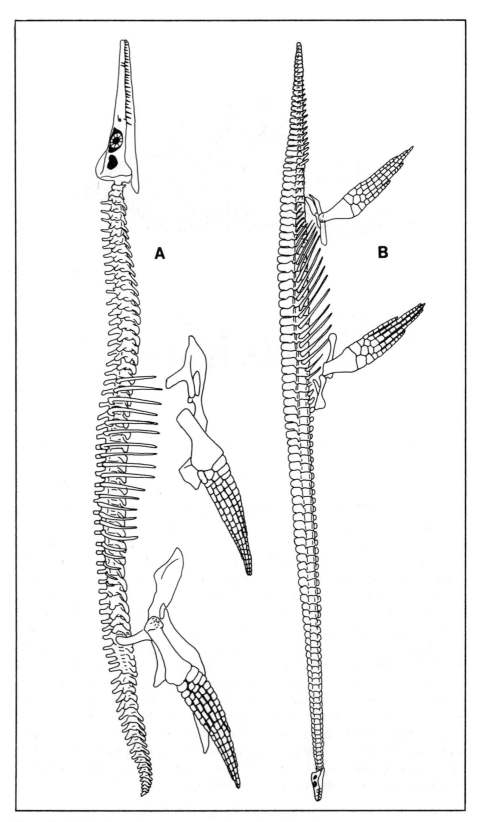

Figure 87—Plesiosaurs. A, the short-necked *Dolichorhynochops*. B, the long-necked *Styxosaurus*. Both from the Kansas Cretaceous.

Dinosaurs.—Dinosaurs are very rare in Kansas. This is because upland deposits of Mesozoic age are rare in the state. The two genera of dinosaurs known from Kansas are both thought to have been beach-dwellers that may have fed in the water. One genus, *Claosaurus,* was a duck-billed dinosaur that grew to as much as ten feet tall. Despite their name, these dinosaurs did not usually dabble in ponds. Thousands of grinding teeth permitted them to eat tough pine needles and other nearly indigestible plant parts. *Claosaurus,* however, may have dined partly on aquatic vegetation.

Silvisaurus was an armored dinosaur that was as much as eight feet long. Generally dinosaurs of this group possessed heavy, knobbed armor all over the upper body parts. Spikes projected from the shoulder region and along the sides of the body. The tail was usually clubbed. However, in *Silvisaurus* the armor seems to have been reduced and lightened. The tail club was absent. This may have enabled *Silvisaurus* to feed on seaweeds or other aquatic vegetation without sinking beneath the water.

Pterosaurs.—Pterosaurs were the only group of reptiles with the ability to fly. In Kansas, they occur only in deposits of the Cretaceous seas (Figure 90).

Pterosaurs are not closely related to either birds or bats. Birds fly by using wing feathers. Bat wings are constructed of hair-covered skin stretched between several elongated finger-bones. Pterosaurs did not possess feathers, and although they probably had hair-covered wing membranes, the wing support was formed from a single finger, the equivalent of your "pinkie."

Pterosaurs had keeled breastbones like modern birds and thus probably had the muscular power to flap if necessary. However, paleontologists believe that pterosaurs specialized in soaring. They used rising air currents to help prolong their flight time. They flew from shore to forage for fish, as do the albatrosses today. Some (from Texas) had wingspreads as large as a commercial airliner, although specimens so far discovered from Kansas have wingspans of less than twenty feet.

In the Cretaceous, most pterosaurs had long, toothless beaks. A pterosaur could swoop down on schooling fish and scoop or pluck prey swimming near the water surface. It is not known whether a pterosaur falling into the water could become airborne again.

Birds.—A unique variety of toothed birds is found in Cretaceous rocks of Kansas. The remains are found as partial or whole skeletons in slabs. Tertiary bird fossils are obtained by screening for individual bones. All bird remains are rare and scientifically very valuable.

Two genera are representative of the toothed Cretaceous birds (Figure 91). The first is *Hesperornis,* a large, seal-like bird whose wing bones were so small that it could not fly. *Hesperornis* lived in the Cretaceous seas and hunted fish, snapping them up in its long, toothed beak. Its flexible neck

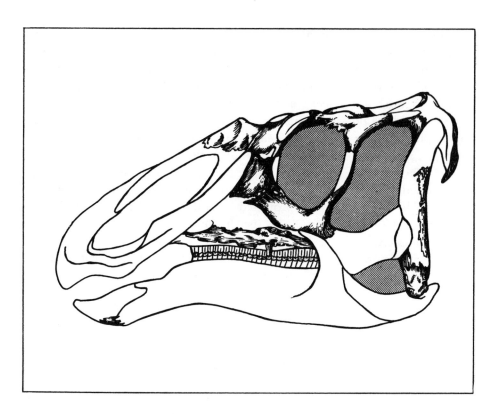

Figure 88—Kansas dinosaurs are rare. This is what the skull of the only Kansas duck-billed dinosaur, *Claosaurus,* might have looked like.

could fold up and extend like a loon's. Its legs bore lobed feet that were folded tightly during the forward phase of each paddling stroke. During the backward power phase, they unfolded like umbrellas. The bird was thus a powerful swimmer and could dive to the bottom to chase prey. The bird's leg bones were jointed in such a way that the legs moved along the side of the body, stroking almost like oars. The legs were located far toward the back of the bird's torpedo-shaped body; because of this, the bird could not walk on land. Incorrect early reconstructions of this bird, still often seen, show it sitting upright. Actually, the bird could progress on land only by flopping along on its belly. *Hesperornis* had to come ashore to lay its eggs, but at these times it must have been very vulnerable to predation.

A second toothed bird of the Kansas Cretaceous is *Ichthyornis.* The toothed beak of this bird somewhat resembles the jaws of a baby mosasaur, but the body possessed pigeonlike wings and a large, keeled breastbone. This bird was a good flyer, and paleontologists believe it led a ternlike existence. It did not often venture very far from shore. Like *Hesperornis,* this bird also preyed upon fish.

Bird remains recovered from Tertiary deposits in Kansas are close relatives of modern forms and belong to the stork, heron, duck, hawk, vulture, and quail families.

Figure 89—This reconstruction of the life appearance of *Silvisaurus* was produced after research by Kansas paleontologist T. H. Eaton and drawn by Kay Reinhardt.

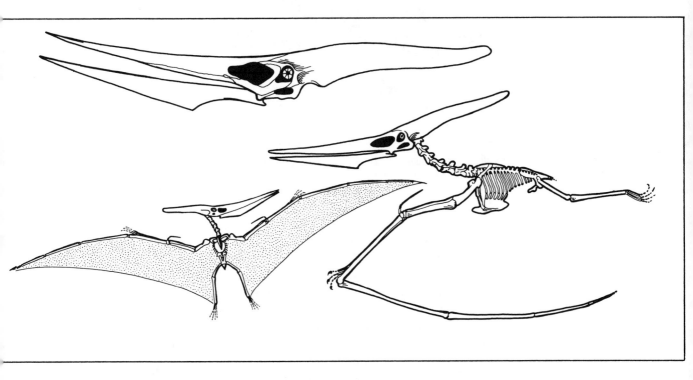

Figure 90—*Pteranodon,* an extinct flying reptile, is known afrom the Kansas Cretaceous. Its long, crested skull is toothless, and its wingspan is up to twenty feet. Stippling shows where paleontologists think the wing membrane was placed in life.

Figure 91—Toothed birds of the Kansas Cretaceous. A, the flightless *Hesperornis.* It could swim and dive well but flopped about like a seal on land. B, *Ichthyornis,* a flying bird with a ternlike lifestyle. Reconstructions based on research by Kansas paleontologists L. D. Martin and J. D. Stewart.

Small Mammals.—Mammal fossils generally are identified from teeth and foot bones. Their tiny remains are recovered by screening loose Tertiary and Pleistocene sediment. A rich deposit of such sediment will yield a shoebox full of bones in about a hundred tons of unsifted sand. The effort required to recover these bones is more than compensated for by their scientific value. Small mammals are sensitive climatic indicators and can give detailed information about conditions in the past. Below is a sampling of common small fossil mammals of Kansas.

Insectivores include the common shrews and moles (Figure 92). Their tiny jaws have sharp-cusped teeth. The upper arm bones (humeri) of moles, designed to help the animal burrow in soft sand, are easy to identify.

Bats are rare in Kansas. Whole skeletons or limb bones are rarely found, but the teeth and jaws are more durable (Figure 93). The picture shows that bat teeth are somewhat similar to mole teeth.

Rabbit bones are fairly common. The lightly constructed skulls are rare, but jawbones with teeth and limb bones are useful. Rabbits have enlarged front teeth for gnawing. A wide space (diastema) separates the front teeth (incisors) from the grinding cheek teeth.

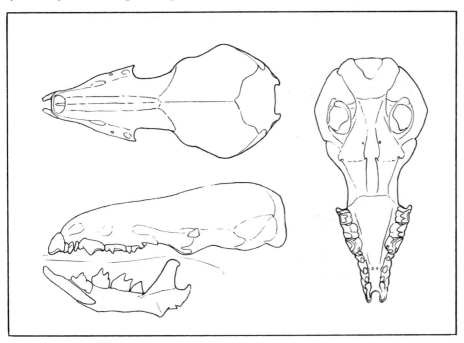

Figure 92—A shrew skull, showing the bladelike teeth with W-shaped enamel pattern. The whole skull is less than an inch long. (Reproduced by permission of E. R. Hall.)

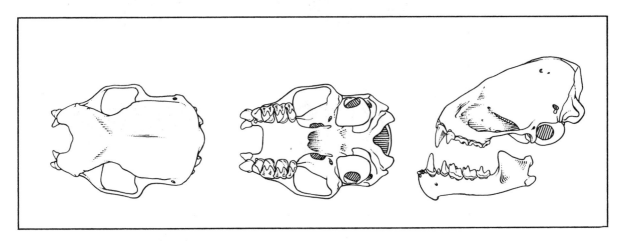

Figure 93—A bat skull, showing the high-cusped teeth with W-shaped enamel pattern. The whole skull is less than an inch long. (Reproduced by permission of E. R. Hall.)

Rodent bones are the most common mammalian microfossils. Like rabbits, rodents have gnawing front teeth and a diastema. Rodent cheek teeth are very diverse in form, and initial indentification is best made by noting that they are small and do not resemble insectivore, bat, or rabbit teeth. Some rodents—notably the porcupines and beavers—grow large, but small rodents known from Kansas are grouped in the squirrel, pocket mouse, hamster, and jumping mouse families.

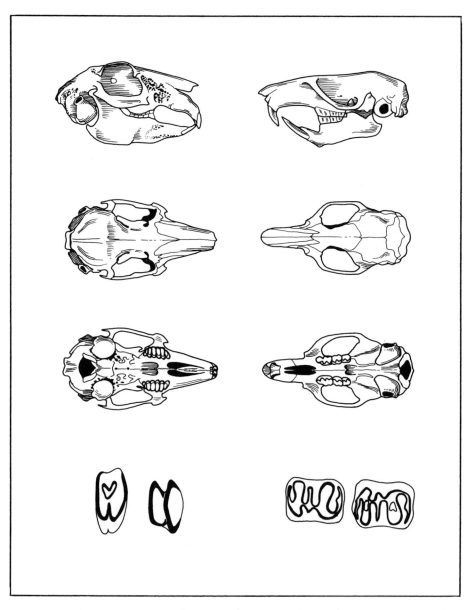

Figure 94—How to tell rabbits from rodents. The rabbit skull, to the left, shows many small perforations in the jaws and skull bones. The figure-eight shaped enamel pattern of the teeth is distinctive. Rabbits have two pairs of upper incisors, while rodents have but one pair.

Figure 95—Mammals move in different ways, according to the structure of their limbs. Those that touch the entire sole of the foot to the ground (as the shrew at top) are called plantigrade. Those that touch the toes to the ground (like the dog at center) are digitigrade. Those that travel only on the toe-tips (as the horse at bottom) are unguligrade. Arrow shows cannon bone of the ungulate.

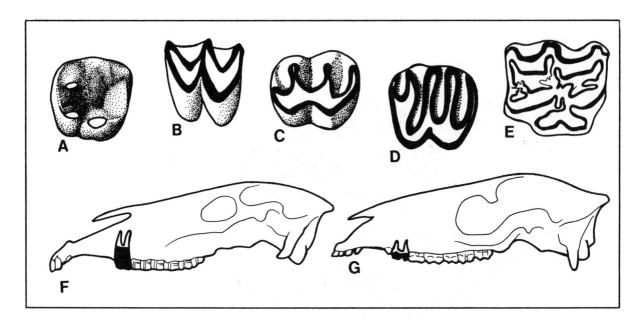

Figure 96—The enamel patterns, shapes, and growth habits of mammal teeth help paleontologists to classify them. A, a bunodont tooth, showing prominent cusps. B, a selenodont tooth, showing sharp, crescentic ridges. C, a lopho-selenodont or combination pattern, showing both loops and ridges. D, a lophodont tooth, showing loops. E, a complex lophodont tooth (of the horse, *Equus*). F, cut-open view of a horse skull, showing a tall-crowned or hyposodont tooth. Modern and other advanced horses have hypsodont teeth. G, cut-open view of a primitive horse skull, showing a short-crowned or brachydont tooth.

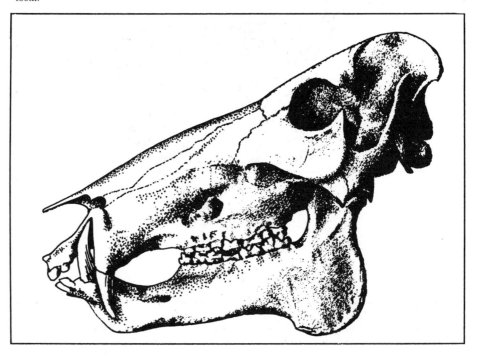

Figure 97—Peccaries are artiodactyls. *Platygonus* is a common Kansas Pleistocene find.

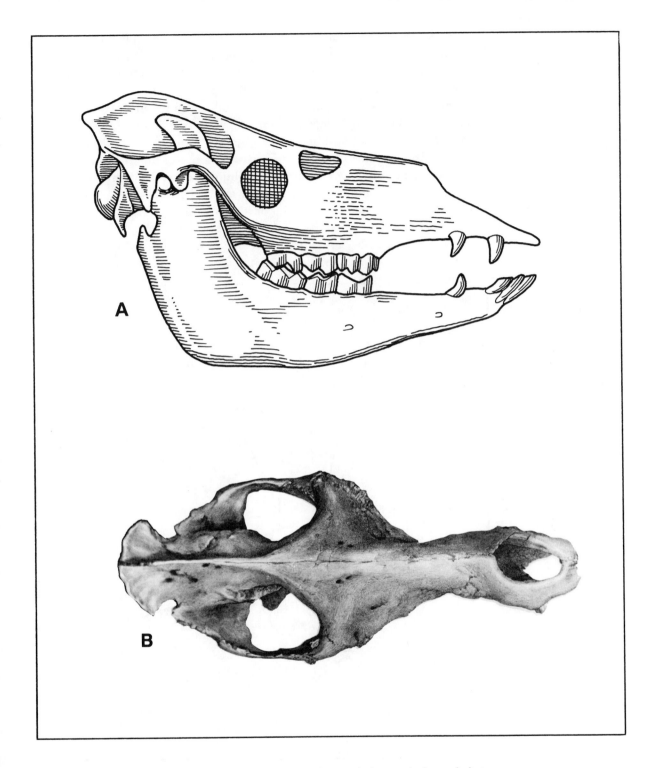

Figure 98—Camel skulls. Camels are large artiodactyls that lack upper incisor or nipping teeth, but appear to have two upper canine teeth (A). B, top view of *Gigantocamelus* skull, from the Kansas Pleistocene. The skull is almost three feet long.

Artiodactyls.—These are hoofed mammals with two to four hoofs per leg. They are commonly referred to as cloven-hoofed animals. Most of them are as big or bigger than a modern Shetland pony, and their remains are found in Tertiary and Pleistocene deposits of Kansas.

Artiodactyl cheek teeth may have low bumps (bunodont) or sharp crests (selenodont). Their limbs may be elongate, with a single fused cannon bone below the knee (wrist) and hock (ankle) joints. Primitive artiodactyls have no cannon bone, but rather several separate bones in this region.

Peccaries are piglike artiodactyls with bunodont teeth (Figure 97). They lack cannon bones and have two toes per foot. Their jaws have both upper and lower incisors, and their canine teeth grow into long, sharp tusks.

Camels no longer live in North America, but they evolved here and were common in the Tertiary and Pleistocene of Kansas (Figure 98). Many kinds are very large; none is smaller than the living llama. They have broad

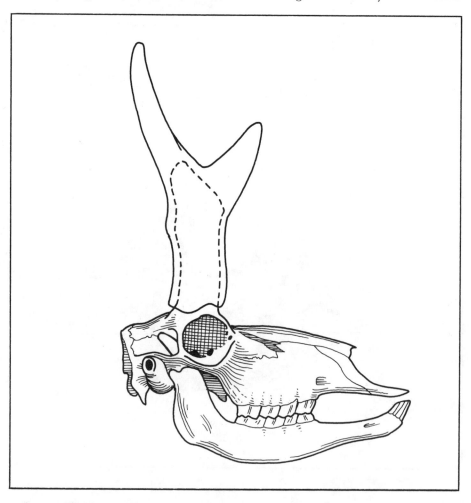

Figure 99—Kansas is the land where the deer and the "antelope" play—and they have done so here for millenia. This is the fossil pronghorn *Proantilocapra.*

Figure 100—Cervid antlers and limb bones are common Kansas Pleistocene remains. Skulls, such as this one of the deer *Odocoileus,* are rare. (Reproduced by permission of E. R. Hall.)

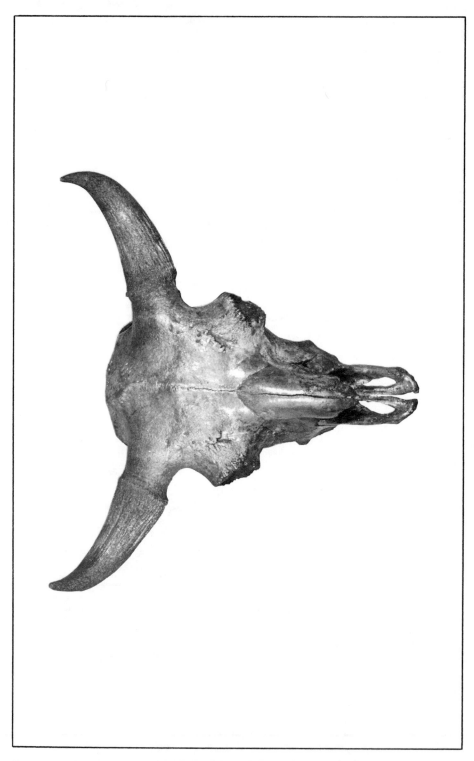

Figure 101—Fossil *Bison* skull recovered from Kansas Pleistocene deposits. Distance between horn tips about two and one-half feet. (Photo courtesy Carl D. Frailey.)

cannon bones, long legs, and spreading toes. Their teeth are selenodont. A long diastema separates front and cheek teeth in camels. Kansas camels lack upper incisors and have several caninelike teeth in both upper and lower jaws.

Pronghorns are nearly extinct. Their one living representative, *Antilocapra americana,* the so-called American antelope, is present in western Kansas. Its extinct relatives were once common all over the state (Figure 99). Pronghorns have long, slender limbs and are all as small or smaller than the living *A. americana.* Their teeth are selenodont and very high crowned (hypsodont). This is because pronghorns eat abrasive grasses and need teeth that do not quickly wear down. They have slender, branching horns that grow upright from above the eye sockets.

Elk, moose, and deer belong to the group called Cervids. They are large mammals having cannon bones and long, sometimes heavy, limbs. Their teeth are selenodont. Their antlers are usually large, complex, and branching, and are shed annually.

Bovids are a varied group, including the true, or African, antelopes, bison, musk oxen, goats, sheep, and some extinct forms. They have short cannon bones, and their teeth are selenodont. Bovid horns are not shed and are never branching. They may grow from the sides or top of the head (Figure 101).

Perissodactyls.—This group includes tapirs, horses, rhinoceroses, and several extinct groups. Only horses and rhinos have been found in the Tertiary and Pleistocene deposits of Kansas.

Perissodactyls have either four, three, or one hoofed toe per foot. The toe corresponding to your middle finger is always largest. Another way to tell perissodactyls from artiodactyls is to look at their ankle bones. Those of perissodactyls have oblique ridges.

Rhinoceroses have four toes per foot. The most common Kansas rhino is *Teleoceras,* a short-legged, barrel-bodied form (Figure 104). Unlike most rhinos, *Teleoceras* lacked a nasal horn. Rhinoceroses have large, square teeth with complex enamel loops (lophodont). *Teleoceras* was a grass-eating rhino; other Kansas rhinos ate leaves and bark. Rhinoceroses became extinct in North America at the end of the Tertiary.

Horses were once much more abundant and diverse than they are today. At the beginning of the Miocene epoch in Kansas (15 million years ago), there were at least ten genera of horses living here, with several dozen species. All but one of these forms specialized in eating grass. The grass-eating forms all have high-crowned teeth. Grass is very abrasive food; hypsodonty allows the animal a longer life-span by prolonging tooth life.

Horses have a single, rounded cannon bone below the knee and hock in each leg. The legs terminate in a single large hoof. Horse skulls can easily be told from similar-looking artiodactyl skulls because horse skulls have upper incisors, while camels, antelopes, cervids, and bovids lack them. Like rhinoceros teeth, horse teeth have lophodont enamel structure.

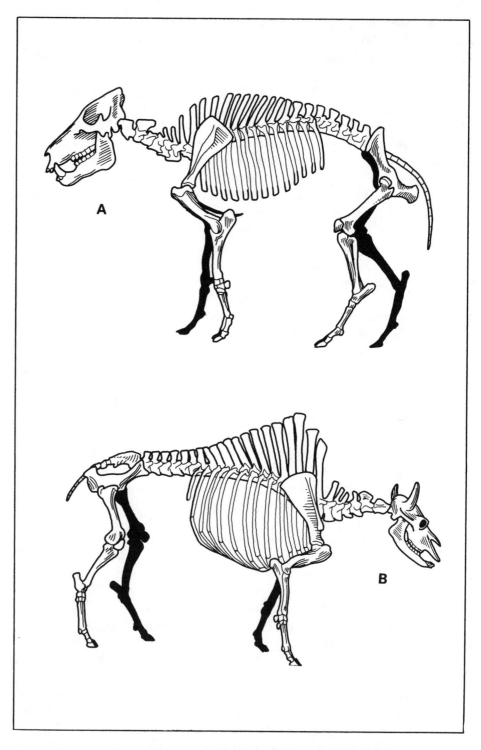

Figure 102—Artiodactyl skeletons. A, *Platygonus,* showing short, unfused cannon bones. B, *Bison.* *Platygonus* stood about two and one-half feet high at the shoulder; some Pleistocene *Bison* measured eight feet at this point.

Proboscideans.—This group includes gomphotheres, mammoths, and mastodons. All three forms occur in Tertiary and Pleistocene deposits in Kansas.

Proboscideans have large but delicate skulls with bubbly looking internal structure. A proboscidean skull is usually quite short from back to front. The trunk is rooted high up on the animal's face and the single large nasal opening looks like an eye socket in the middle of the animal's forehead. The actual eye sockets are small and are located on each side.

Proboscideans have very large, very heavy skeletons. They have four toes per foot and have several separate bones instead of a cannon bone. All have only a single pair of upper incisors, which are enlarged enormously to form tusks.

Gomphotheres have cone-cusped (bunodont) teeth (Figure 107). The cusps are in two rows, one on the outside and one on the inside of each tooth. There are never more than fifteen cusps in a row on a gomphothere tooth. Gomphotheres have large lower tusks that are usually flattened and scoop-shaped. Their lower jaws are long, to hold the large roots of these teeth. Their upper tusks are small and nearly straight. Gomphotheres became extinct at the end of the Tertiary.

Mastodons lived in Pleistocene forests of Kansas. They ate leaves and bark. Most have short, tuskless lower jaws, but large, heavy, curving upper tusks. They stood taller than gomphotheres but not as tall as the mammoths. Mastodon teeth are bunodont like those of gomphotheres, but they possessed at least fifteen cusps on each tooth row; the cusps were smaller than those on gomphothere teeth. These forms are relatively rare in Kansas.

Mammoths are the largest proboscideans. They specialized in eating grass. Like other grass-eating mammals, their teeth are relatively high crowned. Hundreds of tiny cusps in mammoth teeth have fused together into narrow, ridged loops, so that these teeth are no longer bunodont but lophodont. Like mastodons, mammoths have short, tuskless lower jaws but large, heavy, curving upper tusks.

Carnivores.—This group today includes cats, dogs, civets and their relatives (viverrids), bears, raccoons, and badgers and their relatives (mustelids). Only mustelids, certain types of dogs, and cats are commonly found in Kansas fossil deposits. Their remains in Kansas are Tertiary and Pleistocene in age.

Bone-crushing dogs, which are only distantly related to living wolves and coyotes, possessed very heavy skulls and jaws and large, broad teeth. They led a hyenalike existence as scavengers and were capable of crushing the heaviest leg bones to get at the marrow inside. They were long-legged, however, and could probably also hunt in packs.

Mustelids include not only badgers but also skunks, weasels, otters, minks, and some extinct forms. Kansas has produced fossils belonging to all of these groups (Figure 111). *Leptarctus* is a small, extinct form that possessed exceptionally powerful jaw muscles. It may have lived much as does the wolverine of today, digging burrows and aggressively snapping up prey.

Cats come in three varieties: saber-toothed, dirk-toothed, and conical-toothed (Figure 112). Apparently, only the conical-toothed varieties survive today. *Adelphailurus* is a pumalike, conical-toothed cat from Kansas. Conical-toothed cats are generally fleet and agile pursuit-hunters that try to bite the spine for the kill.

Saber-toothed cats are represented in Kansas by *Machairodus*. This lion-sized cat could pursue prey over short distances, much as the living lion. Its short, coarsely serrated canines were used to slash the jugular vein, while the powerful clawed forelimbs grasped the forequarters and pressed back the neck of the prey. These cats probably hunted camels, peccaries, and bovids in Kansas.

Dirk-toothed cats have huge, curving upper canines. *Barbourofelis* is a good example of such a cat. They typically had short but extremely powerful limbs. They were unable to pursue prey, but rather lay in ambush. Gomphotheres may have been a favorite food item. The jaws of these cats could swing open extraordinarily wide to permit free use of the upper canines. The canines themselves were very thin and finely serrated on both front and back surfaces. *Barbourofelis* twisted the neck of prey much as *Machairodus,* but rather than slashing with the canines, carefully inserted them up to the hilt and drew them out again in a curving arc. This permitted *Barbourofelis* to penetrate the thick hide of a proboscidean and sever the jugular vein or carotid arteries buried deep within the neck.

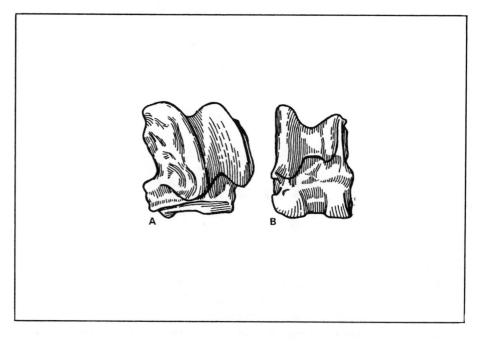

Figure 103—Ankle bones of (A) perissodactyl and (B) artiodactyl. Those of perissodactyls have oblique ridges.

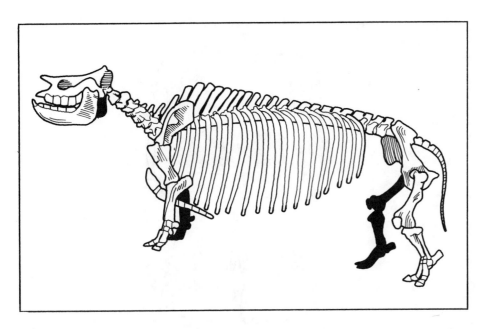

Figure 104—Hundreds of skeletons of the barrel-bodied, short-legged rhinoceros *Teleoceras* have been found in Kansas.

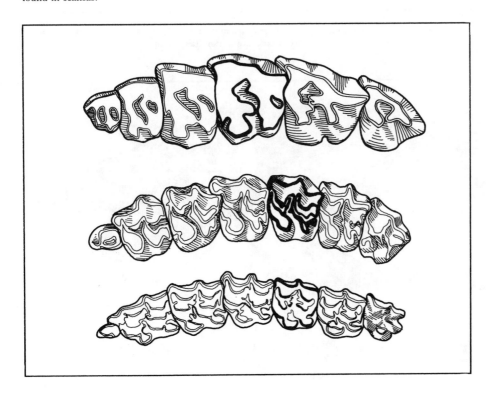

Figure 105—Perissodactyl teeth, commonly found in Kansas Tertiary deposits. Top, rhinoceros. Middle, primitive horse. Bottom, advanced horse. Black emphasizes lophodont enamel pattern on chewing surfaces of teeth.

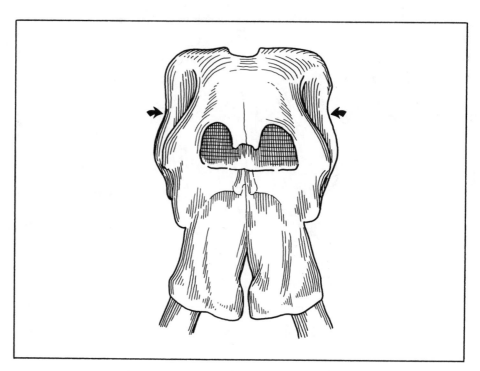

Figure 106—Origin of the cyclops legend? The big hole in the middle of this mammoth skull is the nasal opening at the base of the trunk. The eye sockets are to the sides, as indicated by arrows.

Figure 107—A reconstruction of the extinct Kansas gomphothere *Platybelodon,* based on research by Debra Bennett.

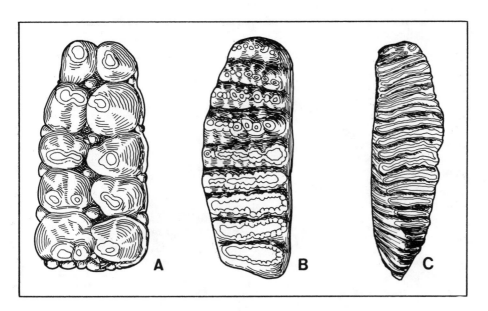

Figure 108—Proboscidean teeth. A,B, gomphothere. C, a mammoth.

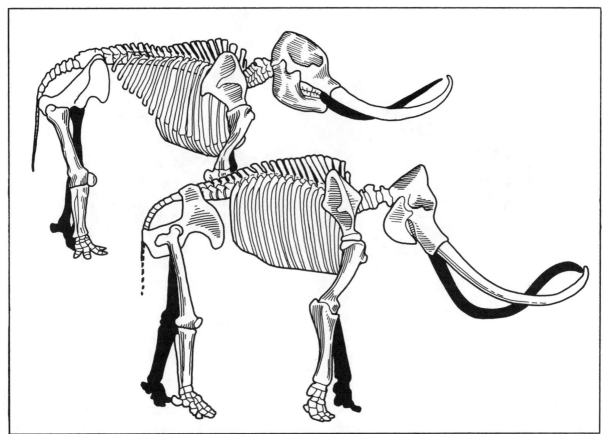

Figure 109—Skeletons of proboscideans. Top, a mastodon. Bottom, a mammoth.

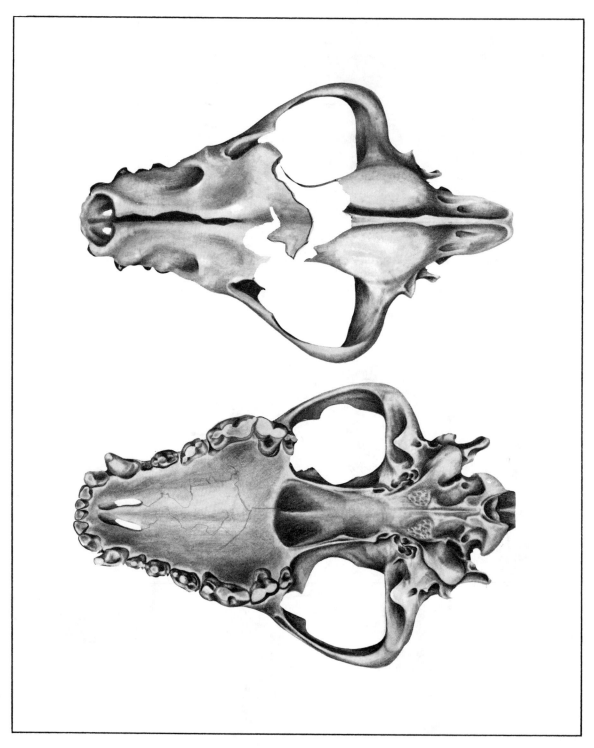

Figure 110—View of the palate (A) and top of the skull (B) of a bone-crushing dog. Notice the heavy bone structure, wide face, and sturdy teeth. Skull length about ten inches. The animal had a hyenalike appearance in life.

Figure 111—Reconstruction of the head of the extinct mustelid *Leptarctus*. This carnivore, about the size of an otter, had extremely strong jaws. Reconstruction based on research by Debra Bennett.

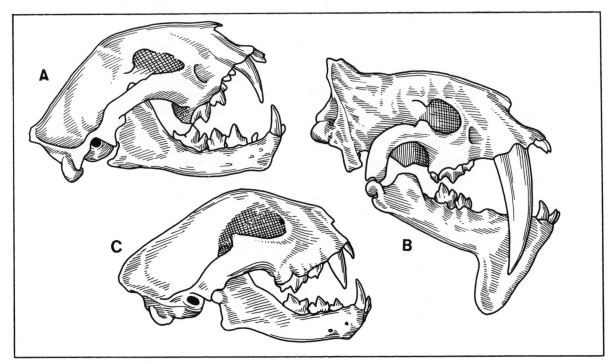

Figure 112—Carnivorous or meat-eating mammals have teeth made for biting and cutting. A, a sabertoothed cat, *Machairodus*. B, a dirk-toothed cat, *Eusmilus*. C, a conical-toothed cat, *Pratifelis*. Cats A and C are known from the Kansas Tertiary and Pleistocene.

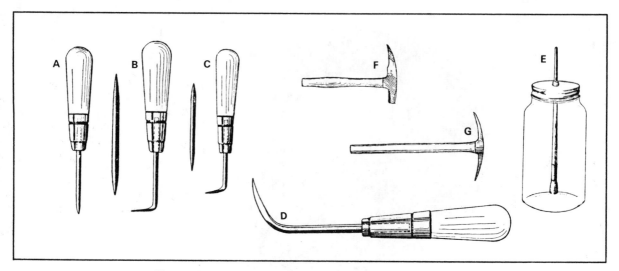

Figure 113—Tools for collecting fossils. A,B,C,D, various small awks, picks, and scrapers. Bent and sharpened screwdrivers also work well. E, jar with brush mounted in screw-on lid, for varnish-based hardening compound. F, geological hammer. A bricklayer's hammer weighing five to seven pounds also works well. G, "marsh pick" or lightweight pick. A U.S. Army entrenching tool also works well. Various small bags, plaster of Paris and a mixing bucket, strips of burlap sacking, old newspapers, and a knapsack are also standard items of field gear.

Collecting Fossils

The accompanying illustration shows good tools to take on fossil-collecting expeditions. You'll be able to collect many fine specimens with these tools, if you remember never to rush when removing fossils from rock or soil. Each fossil has been embedded for thousands or millions of years. Removing it will weaken and sometimes destroy the specimen and must be done with great care. Some fossil specimens are so rare that they have great scientific value. Most kinds of invertebrate fossils and plants are common, although particularly complete or beautiful specimens deserve professional attention. Most kinds of vertebrate fossils are rare. Skulls should always be treated as if they were priceless, for they are the most scientifically useful portion of a fossil vertebrate's skeleton. Judicious use of a varnish-based hardening compound, and correct, professional use of plaster bandages are sometimes the only ways in which delicate skulls, such as those of mammoths and mastodons, can be conserved.

If you should find a fossil that you think a paleontologist should examine, members of the staff at several paleontology museums can help you. Kansas paleontology museums are glad to hear from you and to receive fossils for identification. If your specimens are of scientific value, and you have kept careful note of the location in which they were found, you may wish to donate them for scientific study or exhibit. Only in paleontology museums are facilities available for study and permanent care of these priceless bits of Kansas history.

5

Geologic Log of U.S. Interstate 70

by Rex Buchanan and James McCauley

For tourists and natives, one of the most common routes in Kansas is U.S. Interstate 70, 424 miles of four-lane highway running east and west across the state. From Wyandotte County to Sherman County, I-70 takes in only a fraction of the state. Still, that part of Kansas visible from the Interstate tells a great deal about the state's geology, topography, and history.

A drive along I-70, for example, makes it apparent that Kansas has many transitions in the landscape. In the east, Kansas is hilly and forested, with a number of large lakes and rivers; to the west is an area of more arid, grass-covered plains. Driving all the way across the state on I-70, travelers pass through more than 320 million years of geologic history, from the older Pennsylvanian-age rocks of eastern Kansas to the much younger sediments in the west.

Much of this rock sequence was formed on the bottoms of ancient sea basins that slowly subsided and were filled with layer after layer of sediment. Pressure from deep burial transformed these sediments into sedimentary rocks, and subsequent gentle uplift by geologic forces brought these rocks to their present position. Intense mountain building has not affected the present surface rocks of Kansas, thus preserving the "layer cake" arrangement of the original sediments. Since emerging from below the seas, the rocks have been attacked by the unrelenting process of weathering and erosion. Fairly late in geologic time, the uplift of the Rocky Mountains affected much of western Kansas, making it up to three thousand feet higher than eastern Kansas. The Rockies also nurtured eastward-flowing streams, some of which dried up on the plains of western Kansas, depositing their sedimentary load and burying the older marine rocks. Other streams persisted and continued their erosive work across the state.

As the result of gentle uplift and long periods of erosion, the expression of Kansas geology is very much like a broad set of stair steps leading downward from west to east across the state. These steps are formed by different rates of erosion on the diverse rock sequences. Hard formations such as limestone and sandstone, which resist erosion, form the tops of these stair steps; softer, more erodable rocks such as siltstones and shales make up the intervening slopes. These steps not only take travelers downhill as they drive east across Kansas but also lead them backward through geologic time, since each step down is a step onto rocks of an older age.

The following log is designed to guide travelers through that geology, to help them notice and understand geologic and historic features that are visible along and near Interstate 70. For example, the log labels many of the rock formations that the road cuts through. Geologists have named each of the rock layers in these road-cuts; the log describes the layers and tells something about their unusual features. It also labels the creeks and rivers that I-70 crosses, and describes many of the towns, hills, and other features that are nearby. The log is not limited, by the way, to those areas visible from I-70. Often just a few miles from the Interstate is a vast change in the geology and landscape, easily visible with a few minutes' detour. This log is designed to alert travelers to those areas and to guide them to sights that they might otherwise miss.

To use this log, keep several things in mind. First, the number in the left-hand column corresponds to milepost markers—small, green, numbered signs that line the highway. The log begins at milepost 0, which marks the Kansas-Colorado border, and then moves from west to east across Kansas. Milepost marker 84, for example, is 84 miles east of the state line. The only exceptions to this rule come in eastern Kansas, where I-70 joins the Kansas Turnpike. While the numbering system changes at Topeka and Kansas City, the milepost numbers always grow progressively larger going from west to east. Once past the turnpike's eastern terminal in Kansas City, however, the numbering system returns to the one used along I-70 from the Colorado state line to Topeka.

While the log is written from west to east, it can also be used to locate features while you are driving from east to west. Simply locate the appropriate milepost, then read the log from bottom to top. For example, traveling west from Topeka, you could begin with milepost 361, showing the Statehouse. Then, instead of reading down, simply move up the page to 358.3, which describes the Kansas River north of the highway. Keep in mind also that the log often pinpoints features by using a fraction of a mile. To find these locations, use your car's odometer to gauge the distance between mileposts. Remember, however, that the log reads from west to east. A location at 198.8, for example, would be 0.8 miles east of milepost 198. Traveling west, that location would be 0.2 miles west of milepost 199. Also remember that the mileage numbers are approximate—particularly for features that line the road for long distances—and the road log number may not agree exactly with your car's odometer.

While it is possible to follow the log completely across Kansas, it is just as useful for shorter jaunts on I-70. You need not begin at Kansas City or at the Kansas-Colorado border. If you join the Interstate at Salina, say, you need only locate the appropriate milepost marker and begin identifying the sights as they appear. The log is not only useful for locating geologic and historic sights, by the way. You can also use it to compute the distance to your destination or the mileage to the nearest rest area. If, for example, you are at Hays and are driving east to Wilson, you need only subtract 159 from 206 to find that you have 47 miles to go.

One final note. Emergency stops only are permitted along Interstate 70, and it is not recommended that you try to collect rock or fossil samples at the sites listed in the log. Using this log will not make the miles any shorter, but it may increase your appreciation of some of the natural and historic resources of Kansas.

I-70 Road Log

Mile Post

HIGH PLAINS

0   *Colorado-Kansas border.* Kansas measures 411 miles from east to west along its southern border. The trip across Kansas on I-70 is slightly longer—424 miles—because the Interstate does not travel directly across the state. Today's state line has not always marked the western edge of Kansas, however. In the days of the Kansas Territory, the border extended all the way to the Continental Divide in Colorado, including Denver and much of the Rocky Mountains. When Kansas became a state in 1861, organizers lopped off the western portion of the Territory.

Although the Rocky Mountains begin 150 miles west of the Kansas state line, their presence has had a major effect on the geology and climate of western Kansas. Moisture-laden air traveling across the Rockies is lifted to high elevations and cooled, causing the moisture to condense and fall as rain and snow over the mountains. This air then enters the plains of eastern Colorado and western Kansas much drier than before, creating a "rain shadow" of low precipitation over the High Plains. Rainfall in Kanorado averages less than eighteen inches per year. Precipitation gradually increases across the state to more than thirty-five inches per year in Kansas City. The windiness, low humidity, and abundant sunshine of western Kansas further reduce the effectiveness of the small amount of precipitation that does fall, creating a semiarid climate. Crops can be grown only by use of irrigation or dry-land farming techniques, which require leaving fields unplanted for a year to build up moisture in the soil. The natural vegetation of western Kansas reflects the arid climate. Short grasses (which are drought-tolerant) are dominant, trees are scarce, and desert-type plants such as cactus and yucca are common.

Surface water is scarce in western Kansas. Porous soils allow rapid infiltration of the scant precipitation and the streams are intermittent, carrying water only briefly after periods of rain. However, much of western Kansas is underlain by a thick deposit of sand and gravel, called the Ogallala Formation, which was carried out of the Rockies by streams over geologic time. The aridity of the plains caused these streams to dry up and deposit their loads over a wide area from South Dakota to Texas. The Ogallala has acted like a sponge soaking up rainfall for millions of years, storing it below ground and providing abundant ground water in much of the High Plains. Just the opposite is the case in eastern Kansas. Plentiful precipitation has eroded much of the landscape, exposing rocks with poor water-storage properties. Perennial streams and reservoirs are common; however, plentiful and good-quality ground water is found only in the stream deposits of major river valleys.

1   *Kanorado*. A number of Kansas place-names come from combining parts of two names into one. The name Kanorado, of course, comes from the combination of Kansas and Colorado. Explanation of many of the place-names in Kansas can be found in John Rydjord's *Kansas Place-Names,* which was the source for some information used in this log.

1.7   Middle Fork of *Beaver Creek*.

3   *Ogallala Formation* (poorly exposed). The Ogallala Formation is composed mostly of sand, gravel, and silt, although in some locations it is cemented together. The Ogallala covers most of the western third of Kansas, and in some places is as much as 350 feet thick. The Ogallala is particularly important because in much of western Kansas it is a water-bearing rock formation—an aquifer. This water-saturated rock underlies Kansas and parts of seven other states; its average thickness is about a hundred feet. Water in the Ogallala has slowly accumulated over millions of years as rain water seeped into the ground. Today the aquifer is being rapidly depleted in many areas, primarily because of irrigation. Many of the irrigation wells visible along I-70, particularly in Sherman County, tap the Ogallala aquifer.

Figure 114—The view from Mount Sunflower, in western Wallace County. At 4,039 feet above sea level, it is the highest point in Kansas.

3.4	*I-70 reaches its highest elevation in Kansas,* some 3,910 feet. That is 3,150 feet higher than at the eastern border of Kansas. The highest point in Kansas is Mount Sunflower—approximately 4,039 feet above sea level—located about 22 miles south of I-70, near the Colorado border in Wallace County (see Figure 114).
4.5	The elevation at this point is approximately 3,900 feet.
7.5	REST AREA
11.6	South Fork of *Beaver Creek.*
14.5	Middle Fork of *Sappa Creek.* This creek, though often dry, is a tributary of the Republican River. The name comes from the Sioux word meaning black or dark.
18	*Goodland.* Irrigation has changed agriculture in much of western Kansas, and a number of irrigation wells are visible around the town of Goodland. In 1950 there were just 100 irrigation wells in northwestern Kansas; by 1978 there were 3,400 irrigation wells, watering more than 3.5 million acres. In some parts of Sherman County, the area water table dropped an average of three to four feet per year during the 1970s. Throughout western Kansas, travelers will see huge center-pivot irrigation systems that have done much to change agriculture in this part of the state. Center-pivot systems are large sprinklers mounted on wheels; they irrigate circular patches of ground. Because the wheels could roll over uneven terrain, center-pivots eliminated the need for extensive leveling of ground, which is necessary in other types of irrigation. Thus, center-pivots opened up large areas for irrigation.
35.3	*Sherman County–Thomas County line.* Along I-70, this marks the dividing line between the Central and Mountain Time zones. Much of northwestern and west-central Kansas is in the Mountain Time Zone. Both Sherman and Thomas counties, by the way, are named for Civil War generals. General William Sherman was famous for his march to Atlanta; General George Thomas fought with Sherman in the Atlanta campaign.
40.4	South Fork of *Sappa Creek.* This creek parallels I-70 for several miles along the north side of the roadbed.

42.5 *Loess* exposure in quarry south of I-70. Loess (pronounced luss) is a finely ground dust that covers much of Kansas. This wind-blown silt is responsible for much of the rich soil in the state; in some places it is several hundred feet thick. Thick loess exposures can be seen in the Kansas City area (at milepost 417). Geologists theorize that loess was originally carried by streams when glaciers were melting at the end of the Ice Age. As these streams dried up, they dropped their sediment and strong winds whipped up the finer particles into clouds of dust and dispersed them over large areas of the central U.S.

46 Road to *Levant*.

49 REST AREA. Just west of this rest area, on the north side of I-70, is a depression in the ground that occasionally holds water. These features are common in western Kansas and show areas where the ground has subsided slightly. Scientists are unsure of the exact cause of these features, although they suspect solution of soluble rocks, wind erosion, and settling of near-surface silt deposits. Even buffalos have been suspected of causing these depressions because of their habit of wallowing in mud. This accounts for the local term "buffalo wallows" being applied to these unusual depressions.

49.7 *Prairie Dog Creek.* This creek parallels the north side of I-70 for several miles. It was named by John Charles Fremont in his exploration of Kansas in 1843.

54 *Colby.*

57.4 North Fork of the *Solomon River.* Within twenty miles, I-70 crosses the beginnings of both the Solomon and Saline rivers, two major tributaries of the Kansas River. The Interstate crosses the same rivers in central Kansas, after they flow across much of north-central Kansas and grow much larger.

64 South Fork of *Solomon River.*

70 *Oakley* is named after Eliza Oakley Gardner Hoag, the mother of the town's founder. It is the home of the Fick Fossil Museum, which displays a number of fossils taken from the Cretaceous chalk formations of western Kansas. Oakley, by the way, is the county seat of Logan County. Unlike most county seats, which are situated in the central part of their

counties, Oakley is in the far northeast corner of Logan County.

70.5   North Fork of *Saline River*. The Indians called the Saline the salt river, and French explorers later gave it the name of Saline. The name obviously comes from the salt content that the river gains as it drains the north-central part of Kansas. This river also gave its name to Saline County and the city of Salina.

73   REST AREA

74   South Fork of *Saline River*.

74.9   *Logan County–Thomas County line*. Logan County was once named St. John County after one of the state's early governors. That was before the Kansas Legislature decided to change the name to honor Civil War General John A. Logan.

75.6   *Gove County–Logan County line*.

80   *Campus*. This name came not from a college but probably from camps that formed in the area when the railroads were being built. Monument Rocks is about twenty-one miles south of here.

84   *Elevation is three-thousand feet above sea level*. Between Hays and this point on I-70, the highway climbs about a thousand feet over the course of only about seventy-five miles.

86   *Grinnell*.

89.6   North Fork of *Big Creek*.

94   *Grainfield*. At Grainfield it is easy to see why this area is called the High Plains. The ground is flat and seemingly featureless, although the Saline River cuts through the earth just a few miles to the north.

96.5   REST AREA

100   Road to *Park*. This town was originally named Buffalo. It was later called Buffalo Park, then shortened to simply Park.

107.5   *Quinter*. This is also the exit for Castle Rock Road, a county road that runs about fourteen miles south to the chalk formation called Castle Rock. Castle Rock stands alone on the Smoky Hill Valley, while nearby is a heavily eroded outcrop of the Niobrara Formation that makes up both Castle Rock and Monument Rocks in the western part of Gove County. The old Butterfield Stagecoach line ran past Castle Rock, and the ruts are still visible in the grass just north of the formation.

113.1   *Gove County–Trego County line.* Both of these counties are named for Civil War participants: Captain Grenville Gove and Captain Edgar P. Trego.

115   *Collyer exit.* The county road that runs south of Collyer also winds near Castle Rock. Castle Rock visitors should keep in mind that they are on private property.

120   *Voda Road.* Bohemian settlers reportedly founded this town. Upon digging a well and finding good water, they named the place Voda, which means "water."

128   *WaKeeney.* This Kansas place-name was created from two names combined into one. WaKeeney was named to honor both Albert Warren and James Keeney, two Chicago settlers who moved to western Kansas and bought a section of land. According to the U.S. Postal Service, WaKeeney is still the only town in the United States by that name.

132–133   REST AREA

135.4   *State Highway 147 exit.* To the north this highway goes to Ogallah. To the south, Highway 147 goes about thirteen miles south to Cedar Bluff Reservoir, a lake built on the Smoky Hill River by the Bureau of Reclamation in 1951. Chalk bluffs and outcrops are visible along many of the banks of Cedar Bluff.

136   Just south of Ogallah is the *Ogallah oil field.* Since oil was discovered here in 1951, this field has produced more than 14 million barrels of oil. Today the field contains forty-two wells that are producing from the same Pennsylvanian-age rocks that are visible on the surface in eastern Kansas.

138   This is the approximate boundary between Tertiary-age rocks to the west and older, Cretaceous-age rocks to the east. Chalks and chalky shales are common to the east, while the Ogallala Formation is more common to the west. The Cretaceous

period was named for the Latin word "Creta," which means chalk. Limestones and chalks are common deposits of the seas of this age and are found over large parts of the world, including the island of Crete in the Mediterranean. The famous White Cliffs of Dover, England, are thick Cretaceous chalk deposits similar to the Niobrara Chalk that is widespread in western Kansas. The Niobrara Chalk contains famous fossil beds that have yielded skeletons of fish, marine reptiles, and even flying reptiles from the age of dinosaurs.

139.8   *Spring Creek.* According to the U.S. Geological Survey, Spring Creek is the most popular name for rivers and creeks in Kansas.

140.5   *Riga Road.* The town of Riga was named by Volga-Germans who moved westward from Ellis County to Trego County. Riga means "ridge of sand."

141   Four miles south of I-70 is a hill called *Round Mound* (el. 2,323 feet).

143.7   *Ellis County–Trego County line.* Ellis County was named for George Ellis, a Civil War lieutenant who was killed in the Battle of Jenkins' Ferry in Arkansas. Lieutenant Ellis happened to be in the Eldridge House in Lawrence when it was burned by Quantrill's Raiders in 1863. To the north of Ellis County is Rooks County, named for Civil War Private John C. Rooks. Some forty-seven Kansas counties are named for soldiers, but only Rooks County is named for a private.

145   *Ellis.* This town was the boyhood home of Walter Chrysler of automobile fame.

149.5   *Fort Hays Limestone.* This massive limestone bed often contains fossilized clam shells. It marks the base of the Niobrara Chalk Formation.

153   *Yocemento Road.* This town's name reportedly came from I.M. Yost, who owned a cement plant here, and the Spanish word for cement.

159   *Hays.* Named for the fort that was established there, Hays is today a regional trade center for much of western Kansas. On the campus of Fort Hays State University is the Sternberg

Figure 115—This sphinxlike chalk formation north of Monument Rocks in western Gove County toppled in 1986.

Memorial Museum, which includes a fine collection of fossils dug out of the Cretaceous chalk. General George Armstrong Custer was once stationed at Hays, as well as at Fort Riley and Fort Leavenworth, which are also in Kansas.

160.1   *Chetolah Creek.*

162   REST AREA

166   North Fork of *Big Creek.* To the north is the town of Catharine, named for Catherine the Great, an empress of Russia. Of German ancestry, Catherine offered free land in southern Russia to German immigrants. Many Russian-Germans, or Volga-Germans, as they preferred to be called, later immigrated to the U.S., many of them settling in Ellis County.

Figure 116—The Cathedral of the Plains, built of native limestone, in Victoria, Ellis County.

168.6   *Victoria.* Like Catharine, Victoria was named for royalty. Victoria was settled by Englishmen who named the town after their queen. The English were not particularly successful farmers or ranchers, and most of them left Ellis County. Even from I-70, it is possible to see the twin towers of the church of St. Fidelis in Victoria. Also known as the Cathedral of the Plains, this church was built out of 17 million pounds of native Fencepost Limestone between 1908 and 1911 (Figure 116).

170.5   *Mud Creek.* This is one of nineteen Mud Creeks in Kansas.

BLUE HILLS

172.5   North of I-70 here are the *Blue Hills,* of which the Fort Hays escarpment is a part. The Blue Hills are formed by hard Cretaceous limestones such as the Fort Hays. Major east-flowing streams such as the Saline and the Solomon have cut deep valleys through these rocks with many picturesque side canyons. Exposures of bluish-gray shales interbedded with the limestone probably gave the region its name.

173   *Walker.* This town is the site of another attractive prairie church, St. Ann's, which is also built of native limestone (see Figure 117). Walker was the site of a major air field in World War II, and the runways and hangars are still present about two miles north of the Interstate.

174.6   *Walker Creek.*

175   *Ellis County–Russell County line.* From the pumps that line I-70, it is easy to discern that this is oil country. Ellis and Russell counties are regularly ranked first and second, respectively, in oil production among the 105 counties in the state. These two counties are located atop a subsurface geologic feature called the Central Kansas Uplift, a dome of rock that runs from Barber County in south-central Kansas to Norton County in the northwestern part of the state. About one-third of the wildcat wells drilled in Kansas each year are drilled in the Central Kansas Uplift area.

176   *Gorham road.*

178.5   *Witt sinkhole.* A dip in I-70 at this point shows the location of the Witt sink. This subsidence is caused when ground water dissolves away underground salt beds about 1,300 to 1,600 feet beneath the surface. In some places, the ground water washes

Figure 117—St. Ann's Church at Walker, in eastern Ellis County. The church's spire is visible from milepost 173.

out cavities that suddenly collapse. In other cases, such as the Witt sink, the earth subsides more slowly. While there appears to be little chance of a catastrophic collapse, the roadbed is slowly dropping lower each year.

179   *Crawford sinkhole.* This sinkhole creates a small pond and a dip in the highway just below the underpass on the south side of I-70. Even in dry weather, the pond contains water. This sinkhole is forming much like the Witt sink (described above), although its formation may be putting stress on the bridge crossing I-70 (see Figure 118).

180   Extending several miles north and south is the *Gorham oil field.* Discovered in 1926, this is one of the older oil fields in central Kansas. Today it includes nearly 500 pumping oil wells and has produced more than 88 million barrels of oil.

182.8 *Fossil Creek.* This creek, which parallels the south side of I-70 for several miles, was obviously named for the fossil beds found in the area. The town of Russell was originally called Fossil Creek Station. Later this was shortened to Fossil, then finally changed to Russell. One of the main thoroughfares into Russell, though, is still called Fossil Street.

185   *U.S. Highway 281 exit.* Directly south of I-70 is Fossil Lake.

187   REST AREA

189   South of I-70 is the *Hall-Gurney oil field.* This large field extends south into Barton County. First discovered in 1931, it has been responsible for the production of more than 133 million barrels of oil, which is close to twice the average annual production of the entire state of Kansas today. In 1982, Kansas produced 70.5 million barrels of oil, which made Kansas seventh among the leading oil-producing states in the nation. The value of that oil production amounts to several billion dollars annually.

193   *Bunker Hill Road.*

202.9 *Greenhorn Limestone Formation* (poorly exposed). Fenceposts and buildings made of native limestone line the roads in this area. Most of that limestone comes from a thin but widespread and uniform limestone bed at the top of the Greenhorn Limestone called, for obvious reasons, the Fencepost Limestone.

SMOKY HILLS

205   *Russell County–Ellsworth County line.*

206   *State Highway 232 exit* to Wilson and Lucas. A mile south of I-70 is Wilson, a town that was settled primarily by Czechoslovakians and calls itself the Czech capital of Kansas. North of I-70 about seven miles is Wilson Lake, a man-made lake on the Saline River. The Wilson Lake area offers deep canyons and steep scenic hills that belie the notion that western Kansas is flat and featureless. Farther north, in the town of Lucas, is the Garden of Eden, a series of unique cement statues built by a disabled Civil War veteran, S. P. Dinsmoor.

208   *Dakota Sandstone.* Just north of I-70 are several unusual erosional features composed of Dakota Sandstone, a Cretaceous-age sandstone that was probably deposited near the shore of a Cretaceous sea. These features are not easily visible from the east-bound lane. The Cretaceous seas were among the most

Figure 118—An area of slow subsidence at milepost 179 along I-70, in Russell County.

widespread in all geologic history and covered much of North America. The Dakota Formation occurs throughout the north and central Great Plains. From here on west it is in the subsurface of western Kansas and eastern Colorado, coming to the surface as a sharp sandstone ridge just east of the Colorado Front Range.

209–212   *Wilson channel.* At this point, I-70 passes between the Smoky Hill and Saline rivers. However, geologists have discovered that at about this point the Saline River once meandered south and joined the Smoky Hill. This flat, valleylike area represents that old stream channel.

211.2   *Cow Creek.*

214.6   *Buffalo Creek.*

219   *State Highway 14 exit.* This road runs seven miles south to Ellsworth, a former cow town. James Butler (Wild Bill) Hickock, one of the town's early residents, later moved to Hays and became town marshal. And Wyatt Earp reportedly got his start as a law officer when he made two arrests in Ellsworth.

220.4   *Fencepost Limestone* on frontage road north of I-70. Not easily visible from the east-bound lane.

222   *State Highway 14 exit.* This highway goes about thirteen miles north to Lincoln, the county seat of Lincoln County. Many of the town's buildings are made out of native Fencepost Limestone, and the road north of Lincoln is lined with stone fenceposts.

224   REST AREA. The rest area on the east-bound lane contains several mushroom-shaped sandstone concretions that are found throughout the area. Similar (but much larger) collections of these concretions are on view at Mushroom Rocks State Park north of Kanopolis Reservoir and at Rock City southwest of Minneapolis.

225.7   *East Elkhorn Creek.* Dakota Sandstone outcrops are present on the hill south of I-70.

228.2   *Ellsworth County–Lincoln County line.* At this point there is an outcrop of the Graneros Shale overlain by the Greenhorn

Limestone. Because shale is easily eroded, shale outcrops often seem to only color the road-cut rather than stand out as prominently as a bed of limestone or sandstone.

229   *The Smoky Hills* dominate the view to the east. These hills got their name because they are often obscured by atmospheric refraction and haze, particularly in the draws and river valleys.

233.1   *Dakota Formation* on the north side of I-70.

233.5   *Carneiro Exit.* Carnero is the Spanish word for sheep, and sheep were once commonly pastured on these Smoky Hills. Today, cattle are the most common livestock.

235.5   *Lincoln County–Saline County line.*

236.7   *Dakota Sandstone* on south side of I-70. The hills on the south side of the road are capped by the Dakota, making them more resistant to erosion.

237   *Mulberry Creek.*

238   *Dakota Formation* on south side of I-70.

238.5   *Brookville exit.* Many of the buildings in this town are built of Dakota Sandstone. Brookville is also the site of the Brookville Hotel, a historic Kansas eatery.

242.9   *Mulberry Creek.*

244.2   *Dakota Sandstone.*

246   *Soldier Cap.* About eleven miles south of I-70 is a flat-topped hill called Soldier Cap (el. 1,578 feet). This hill is located in the Smoky Hill Bombing Range, a site where planes from McConnell Air Force Base in Wichita make practice bombing runs.

247   *Iron Mound,* another Dakota-capped hill, is visible to the southeast.

251   *U.S. Highway 81/Interstate 135 interchange* with I-70.

251.5   *North Pole Mound* (el. 1,466 feet) is visible about five and a half miles north of I-70.

253   *Mulberry Creek.*

254.1   *Saline River.* Shortly after it crosses under I-70, the Saline River joins the Smoky Hill.

258.5   *Iron Mound* (el. 1,497 feet) is visible about seven miles south of I-70. This hill is capped by Dakota Sandstone, allowing it to resist erosion. Iron Mound marks the eastern extent of this Cretaceous sandstone.

259.5   This is the approximate boundary between the Permian and Cretaceous-age formations in Kansas. To the east the rocks are older, often containing limestone and flint. To the west, the most common rocks are the clays, shales, and sandstones of the Dakota Formation. The Permian rocks in Kansas were deposited in a vast shallow sea where the environment was continually changing; thus, diverse sediments were deposited. The rocks found along I-70 are generally alternating beds of limestone and shale. Some of these limestones contain chert, or flint, and are resistant to erosion, forming the Flint Hills to the east. South-central Kansas contains brightly colored, Permian-age red beds composed of shale, silt, sand, and occasional beds of gypsum.

264.3   *Solomon River.* Just south of the town of Solomon the Solomon River joins the Smoky Hill River; the confluence is visible from a county road about one mile south of Solomon. Around this point, there are salt springs near the river, marking an area where Permian-age salt beds are near the surface. Much of central Kansas is underlain by Permian-age salt deposits in the Hutchinson Salt Member of the Wellington Formation. Mined in Reno and Rice counties, this salt bed is several hundred feet thick in places. However, where it approaches the surface, as it does in this area, the salt dissolves away and often contaminates local ground water and surface water supplies. Also at this point, Iron Mound is visible to the southwest.

265.5   REST AREA

266   *Saline County–Dickinson County line.* This boundary is also the location of the Sixth Principal Meridian. This longitudinal line runs through Wichita; in fact, it is the source of the name of Meridian Street in Wichita. The Sixth Principal Meridian is a survey line that was used to lay out the public lands of the early Kansas and Nebraska territories. It marks the division be-

tween eastern-range townships and western-range townships and is used in determining legal descriptions of land in Kansas, Nebraska, Wyoming, Colorado, and part of South Dakota. Its location does not correspond to any important longitudinal meridian but rather marks the point where the original survey of the Kansas-Nebraska line was halted due to fear of hostile Indians.

269–271   *Sand-dune topography.* Recently, in geologic terms, this area was covered by sand dunes. While grass has since covered the hills, the rolling land surface still displays that sand-dune topography. The sand in this area probably accompanied an ancestral version of the Smoky Hill River, much as sand dunes are common today around the Arkansas River in central and western Kansas.

274.2   *Mud Creek.*

275   *Abilene.* Another former cow town, Abilene is today the site of the Eisenhower Museum and Library, one of the leading tourist attractions in Kansas.

280   *Sand-dune topography.* See markers 269–271.

FLINT HILLS

281.8   *Nolans Limestone* on the north side of I-70.

282   *Lone Tree Creek.*

285.8   *Winfield Limestone.*

286.4   *Chapman Creek.* To the north is the small town called Moonlight, which is named after a Civil War veteran and not the lunar reflection, as one might suppose. Thomas Moonlight was later the governor of territorial Wyoming.

289.5   *Geary County–Dickinson County line.*

294   REST AREA. *Goose Creek.*

295   *U.S. Highway 77 exit.* To the north is Milford Reservoir, Kansas' largest man-made lake with over 16,000 acres of water. Just south of I-70 is a stretch of the Smoky Hill that was once named Kansas Falls because of a series of falls in the river.

298.7   *Smoky Hill River.* Shortly after it is crossed by I-70, the Smoky Hill joins the Republican River to create the Kansas River.

299.2   *Florence Limestone.* This massive limestone is common along road-cuts in I-70. It ranges in thickness from twelve to forty-five feet and it often contains a variety of fossils. Florence Limestone is one of the chert- or flint-bearing limestones that gave the Flint Hills their name. The flint appears as bluish-gray beds in the otherwise buff-colored limestone.

299.8   *Franks Creek.*

301–302   *Grant Ridge.* This series of hills on the southeast side of I-70 is capped by the Fort Riley Limestone Member. Also called Military Ridge, these hills overlook the Smoky Hill River valley that is today the Fort Riley Military Reservation.

303.6   *Clarks Creek.*

Figure 119—This road-cut at milepost 305 along I-70 shows alternating layers of limestone and shale. The uppermost layer is the Fort Riley Limestone.

| 305 | *Florence Limestone,* overlain by *Oketo Shale,* overlain by *Fort Riley Limestone.* This road-cut is one of the more dramatic along I-70 (see Figure 119). |
| | |

305   *Florence Limestone,* overlain by *Oketo Shale,* overlain by *Fort Riley Limestone.* This road-cut is one of the more dramatic along I-70 (see Figure 119).

306.1   *Fort Riley Limestone,* obviously named for the nearby fort.

306.8   *The Matfield Shale,* overlain by the *Florence Limestone.* The Matfield Shale is a variety of colors, from bright red to a dull green, and it is common along I-70 road-cuts.

308.2   *McDowell Creek.*

308.6   The *Easly Creek Shale* up through the *Crouse Limestone, Blue Rapids Shale, Funston Limestone,* and the *Speiser Shale.*

309.5   REST AREA. The rest area on the west-bound lane is decorated by several red quartzite boulders brought south by glaciers. Pressee Creek runs nearby.

310.3   *Wreford Limestone.* This formation is characterized by an abundance of chert.

311   *Konza Prairie.* Immediately north of I-70 is an 8,616-acre area of tall-grass prairie that was purchased by the Nature Conservancy and is used for research by Kansas State University. Much of the prairie was originally part of a ranch owned by Chicago industrialist C. P. Dewey; the ground is covered by blue-stem grass that has never been plowed. The Konza Prairie is named after the same Indian tribe that supplied the name for the state of Kansas. Konza is one of more than a hundred variations in the spelling of the name for the Kansa Indians.

311.5   *Florence Limestone.*

312.2   *Wreford Limestone.*

312.7   *Swede Creek,* named for Peter Carlson, who settled in Riley County in 1857.

313.1   *Matfield Shale,* overlain by *Florence Limestone.*

314   *State Highway 177 exit.* At this point, the elevation is 1,438 feet above sea level. This is about the highest point along I-70 in eastern Kansas. From here west the highway begins to drop,

following the Smoky Hill River valley. It does not regain that lost altitude until a point just west of Salina. North on Highway 177 is Manhattan, home of Kansas State University. Farther north is Tuttle Creek Reservoir. South on Highway 177 is Council Grove, a historic spot on the Santa Fe Trail.

314.6   *The Matfield Shale,* overlain by the *Florence Limestone.*

316   *Riley County–Geary County line.*

316.1   *South Branch of Deep Creek.*

316.3   *Eskridge Shale,* overlain by *Beattie Limestone,* overlain by *Stearns Shale,* overlain by *Bader Limestone.*

316.8   *Eskridge Shale,* overlain by *Beattie Limestone.*

318.1   *Neva Limestone* on the south. Also scattered along the south side of I-70 are oil wells in the Yaege oil field. Oil production is relatively uncommon in northeastern Kansas, although there has been more production and exploration in recent years. This oil field was discovered in 1959.

319.7   *Threemile Limestone.* Before it is exposed to erosion, this limestone can be light gray to nearly white. In most places it is a massive, thick layer of rock.

321   About two miles north of the Interstate, partly obscured by hills, are *Tabor Valley* and *Tabor Hill,* named after A. W. Tabor, who later moved to Colorado and discovered gold at Leadville.

321.2   *Schroyer Limestone.*

321.6   *Wreford Limestone.*

322   *Riley County–Wabaunsee County line.*

324–325   At about this point I-70 crosses the *Humboldt fault zone,* a series of faults that runs from Nemaha County in northern Kansas to Sumner County in south-central Kansas. The fault zone is associated with the Nemaha Ridge, a buried mountain range composed of igneous rock. The fault zone was probably responsible for producing one of the largest earthquakes recorded in Kansas. The quake, which occurred in April 1867 near Wamego, was felt as far away as Dubuque, Iowa. In nearby Manhattan people fled to the streets, and several foundations were cracked. Farther west near Solomon, a train shook so much that its engineers abandoned it. Less powerful earthquakes struck the area in 1906 and twice in 1929. While there have been no strong earthquakes in the vicinity since that time, the Kansas Geological Survey has recorded a number of "microearthquakes" in the Wamego area that are too small to be felt, measuring less than 2.5 on the Richter Scale.

326.1   *Wreford Limestone.* This exposure has several thin layers of chert running through the limestone.

327.5   *State Highway 99 exit.* Along this highway, about five miles north of I-70, is a large field of glacial boulders on the side of a hill, probably representing an area where a retreating glacier began to melt and dropped much of its load of rock debris. That debris includes a number of red quartzite boulders— many today stained green by accumulations of lichen—that were transported south from South Dakota or Minnesota, a distance of several hundred miles. From here east, I-70 is usually within a few miles of the farthest advance of the Ice Age glaciers in Kansas, which occurred during the Kansan Glaciation. Glacial deposits are common to the north of this line.

328.7   Northwest of I-70 is a creek valley that may mark the ancestral course of the Kansas River before it took a more northerly route. Today, a railroad runs through this creek valley. Roads and railroads often follow creek and river valleys because they represent level, easy surfaces for travel. Interstate 80 in Nebraska, for example, follows the Platte River much of its way.

333.2   *Mill Creek.* This stream drains parts of northern Wabaunsee County before joining the Kansas River. It is one of at least ten Kansas streams named Mill Creek.

333.6   *Hamlin Shale.*

335   Exit to *Skyline Scenic Drive.* This road runs south of I-70 and rejoins the Interstate at the Wamego-Alma exchange just east of Manhattan. The drive traverses parts of the Flint Hills.

336–337   REST AREA

Figure 120—Buffalo Mound, in Wabaunsee County, as seen from the rest area at milepost 336 along I-70.

339   At this point the road crosses *Buffalo Mound.* At 1,273 feet in elevation, it is one of the most prominent points in Wabaunsee County (see Figure 120). It was reportedly named because its shape resembles a buffalo's back. Geologists consider Buffalo Mound a landmark on the eastern edge of the Flint Hills, and outcrops on the mound are good locations for collecting fusulinid fossils. This point is also an approximate boundary between the Permian- and Pennsylvanian-age formations in Kansas. Along I-70 Pennsylvanian rocks are generally alternating limestones and shales with occasional sandstones and thin coal seams. Thicker, economically important coal beds of Pennsylvanian age are found in southeastern Kansas.

342   *Keene Road exit.* Taken south, this road intersects with State Highway 4, which meanders through the middle of the Flint Hills. Highway 4 passes through Eskridge—a town that bills itself as the Gateway to the Flint Hills—and past a pretty, man-made reservoir called Lake Wabaunsee. Buffalo Mound is clearly visible on the western horizon.

GLACIATED REGION

342.2   *Pillsbury Shale,* overlain by the *Stotler Limestone.* The Pillsbury Shale often contains a thin bed of coal.

344.4   *Dover Limestone.*

346   *Wabaunsee County–Shawnee County line.* A number of the place-names in Kansas came from the Indians. Wabaunsee County was named for a Potawatomi chief whose name meant Dawn of Day.

347.3   *Willard Shale,* overlain by *Tarkio Limestone.* This Pennsylvanian-age limestone is gray but weathers to a deep yellow-brown. It is characterized by an abundance of large fusulinids.

348   *Vassar Creek.*

349   *Zeandale Limestone.* This formation was named for a small town east of Manhattan. *Zea* is Greek for grain, and this part of the Kansas River valley was known as Zeandale Bottoms.

350.8   *Mission Creek.*

351.5   *Blacksmith Creek.* This small creek joins Mission Creek and drains directly into the Kansas River. Here the road also passes through *Hickory Knob.* Along the road is the *Scranton Shale,* overlain by the *Bern Limestone.*

352.8   *Auburn Shale,* overlain by the *Emporia Limestone.*

354.4   *Burlingame Limestone.* Fusulinid fossils and algal remains are common in this limestone layer, which ranges in thickness from one to twenty-five feet.

358.3   Just north of I-70 is a view of the *Kansas River.* The Kansas, or Kaw, drains much of northeastern Kansas, beginning at the confluence of the Republican and Smoky Hill rivers near Junction City and emptying into the Missouri River at Kansas City. Other notable rivers that drain into the Kaw include the Solomon, the Saline, the Delaware, the Blue, and the Wakarusa. In all, the Kaw drains a watershed of more than 60,000 square miles, carrying an average of 4.75 million acre-feet of water per year. In northeastern Kansas, where groundwater supplies can be rare, the Kaw is an important water source. It is no accident that many of the state's major cities—

Topeka, Lawrence, Kansas City—are perched on the banks of the Kaw.

361–363   View of *Kansas Statehouse*. Construction on today's Statehouse began in 1867, using limestone from the Fort Riley Member, quarried near Junction City (outcrops of the Fort Riley can be seen at milepost 306). Construction on the west wing began in 1869 using limestone from the Cottonwood Member, dug near Cottonwood Falls in Chase County. Located on a twenty-acre square in the center of Topeka, the Statehouse features murals by John Steuart Curry on its second floor.

365   Last milepost along nonturnpike I-70. Just east of here, I-70 joins the Kansas Turnpike and takes on a new numbering system, beginning with marker 182.

183   *Topeka Service Area.*

183.7   *Calhoun Shale,* overlain by *Topeka Limestone.*

187.3   *Topeka Limestone.* Named for the Kansas capital, this formation ranges in thickness from thirty-three to fifty-five feet.

188–189   *Oregon Trail crosses I-70.* Kansas was on the route of many trails west in the 1800s. The Santa Fe Trail, for example, extended from northeast Kansas to southwest Kansas and is roughly paralleled by today's U.S. Highway 56. Today's I-70 is not too far from much of the path of the Smoky Hill Trail, the quickest route to the Denver gold fields, which were discovered in 1859. When California gold was discovered in 1848, traffic along the Oregon Trail increased. One branch of the trail ran from Independence, Missouri, to Topeka, then northwest to Marysville, and into Nebraska; its course would have crossed I-70 at about this point. Although wagon ruts are not visible at this location, they can be seen along a number of old wagon trails and stage trails that cross the state.

188.5   *Shawnee County–Douglas County line.* Here I-70 passes over U.S. Highway 40, which was one of the major east-west routes through Kansas before the days of the Interstate highway system.

189.5   *Big Springs Anomaly.* At about this point, I-70 crosses a feature that geologists call the Big Springs Anomaly. According to precise measurements by geophysicists, magnetic levels in this area are significantly higher than in the surrounding countryside. These readings are intriguing because they give clues about the subsurface. For example, the Big Springs Anomaly indicates that the underground igneous rock contains more magnetite than in other areas, producing the higher magnetic readings. Geologists have mapped magnetic levels across Kansas, and they use the information to explore for oil and gas and to speculate about the geologic history of Kansas.

193   *Lecompton* is visible to the northeast. This small town was once the capital of territorial Kansas and was known as a proslavery stronghold. Today, bald eagles often roost in the winter in the trees along the Kansas River near Lecompton. In fact, the town was once called Bald Eagle.

193.9   *Tecumseh Shale,* overlain by *Deer Creek Limestone.* This shale layer is as much as sixty-five feet thick near the Kansas River.

198.1   *Plattsmouth Limestone.* The Oread Formation is made up of four smaller limestone layers. The Plattsmouth Limestone is the thickest of those limestones.

199   *Oread Limestone.* The Oread Limestone is one of the most prominent formations in eastern Kansas, averaging about fifty-two feet in thickness in the northern part of the state. The Oread was named for Mount Oread, the hill that overlooks downtown Lawrence. Mount Oread, home of the University of Kansas, is capped by the Oread Limestone. The hill's name came from the home of Eli Thayer, a Massachusetts resident who was one of the foremost promoters of the New England Emigrant Aid Society, which helped settle Kansas in the 1850s. Mount Oread was the name of Thayer's home; he also operated the Oread Female Seminary near Worcester, Massachusetts.

203   *Kansas River.* The first bridge across the Kansas River at Lawrence was completed in December 1863. Construction was held up when eight members of the building crew were killed in Quantrill's raid. When the wooden bridge was finally finished, the construction company charged a toll of twenty-five cents per trip.

204–205   The Kansas River has wandered across this area for centuries, regularly changing course. To the north are several *old river channels* that the Kaw has abandoned, although they still often hold water. Sandpits appear on the south side of the road, and Mount Oread and the University of Kansas are visible to the southwest.

205.8   *Douglas County–Leavenworth County line.*

206   *Mud Creek.* Blue Mound (el. 1,052 feet) is visible 6.5 miles to the south. This hill has resisted erosion because it is capped by the *Oread Limestone.* Standing more than 250 feet above the nearby Wakarusa River, the hill was once the site of a ski slope. Although Blue Mound is shaped much like Iron Mound (the hill east of Salina seen at milepost 258.5), Blue Mound appears more rounded and less steep because it is covered by deciduous trees rather than short prairie grasses.

207   *Stranger Formation.* Best seen on the north side of the highway, this formation is named for the creek that runs nearby.

209   *Lawrence Service Area.*

210.7   *Nine-Mile Creek.*

212   *Stranger Formation,* overlain by reddish *Pleistocene deposits.* The Pleistocene period is the most recent in geologic history, and these deposits are the rocks, gravels, and silts that were dropped by glaciers, streams, or wind.

212.2   *Cow Creek.*

213.5   On the south side of the highway is an exposure of one of the thin *coal beds* in the *Stranger Formation.* These coals were once mined in this vicinity and in other parts of northeastern Kansas by means of small underground mines. Today, all Kansas coal is mined in surface pits in southeastern Kansas.

214.9   *Stoner Limestone Member of the Stanton Formation.* This limestone is generally ten to twenty feet thick in this area but thickens to fifty feet in Montgomery County in southeastern Kansas.

216   *Stranger Creek.* Stranger is a translation of the Indian word "okeetsha," which means "wandering aimlessly about," an apt description of the meandering habit of this stream.

218.2   To the west on the skyline is the *Oread escarpment,* which is capped by the resistant limestones of the *Oread Formation.* This prominent break in the landscape extends from Doniphan County in the far northeast corner of Kansas to Chautauqua County on the Oklahoma border.

220.7   *Tonganoxie Sandstone* on both sides of the highway. Several exposures of this sandstone will be visible to the east of this point. Note the sandstone's cross-bedded appearance—thin, angling lines in the rock indicating that it was deposited by flowing water. The Tonganoxie occurs in a relatively narrow band, never more than twenty miles across, that extends to the southwest. This sandstone reaches a thickness of 160 feet and appears to be the valley deposit of a large southward-flowing river that cut downward into older rocks during the Pennsylvanian period. This formation is named after a small town in southern Leavenworth County.

221   *Stanton Limestone.*

221.4   *Bonner Springs Shale,* overlain by *Plattsburg Limestone.* The Plattsburg is an average of about twenty-five feet thick in Kansas.

221.8   *Wolf Creek.*

222.3–222.9 *Wyandotte Limestone,* upward to *South Bend Limestone.*

222.4   *Leavenworth County–Wyandotte County line.*

224   *State Highway 7 exit.* To the north are the Agricultural Hall of Fame and the city of Leavenworth, and to the south is Bonner Springs. Leavenworth is the site of historic Fort Leavenworth. Established in 1854, Leavenworth was the childhood home of Buffalo Bill Cody.

224.5   *Toll Plaza.* Shortly east of this point, the milepost numbering system changes back to the one used for I-70 west of Topeka.

415.5   *Argentine Limestone Member* of the Wyandotte Limestone, on north side of the highway.

416   *Loess exposure.*

416.9   Thick *loess deposits* on both sides of the highway.

417   *Loess* on north side of highway.

417.5   *Loess* on north side of highway. While loess is common throughout Kansas, especially thick deposits are found in the northeastern corner of the state. Many of the bluffs overlooking the Missouri River in Doniphan County are composed of loess.

418   *Lane Shale,* overlain by the *Argentine limestone.* The Argentine has been mined extensively in the Kansas City area, producing more than 120 million feet of underground storage space.

418.1   *Kansas River* to the south. Between 1854 and 1866, thirty-four steamboats paddled up the Kaw River and one made it as far as Fort Riley. But shallow water and sandbars eventually halted steamboat traffic. As one Lawrence newspaper editor wrote, the Kansas River was "a hard road to travel."

418.2   *Lane Shale,* overlain by *Argentine limestone.*

418.6   *Lane Shale,* overlain by *Wyandotte Limestone.* The Argentine Limestone Member of the Wyandotte Formation is one of the limestones that have been mined extensively in the vicinity of Kansas City, leaving large areas of underground space. In fact, more than 120 million square feet of underground space exist within twenty-five miles of Kansas City. The constant temperature and humidity of these man-made caves make them ideal for the warehousing of all types of items, including government records. The cool temperature allows economical refrigeration and the cold storage of large amounts of food. The underground space has also been used for offices and factories.

419   *I-635 Interchange.*

420   *Cherryvale Shale,* overlain by *Drum Limestone.*

420.3   *Cherryvale Shale,* overlain by *Drum Limestone.* The Cherryvale Shale is named for the town in Montgomery County. In Oklahoma, this formation is called the Nellie Bly Shale.

420.5   To the south, in the Kansas River valley, is an area of Kansas City, Kansas, called the *Argentine.* The name comes from the Latin word for silver, "argentum." A smelter that once operated in this area actually refined small amounts of silver.

421.5   *At this point, I-70 drops down to the flood plain of the Kansas River* and reaches its lowest point in Kansas, approximately 760 feet above sea level. This is 3,150 feet lower than the highest point in Sherman County. The lowest point in Kansas, by the way, is in Montgomery County, south of Coffeyville, where the Verdigris River enters Oklahoma. Here the elevation is only about 700 feet above sea level.

422.5   Thick *loess deposits* on the west side of the highway. Downtown Kansas City, Missouri, is on the bluffs to the east.

423.5   To the west is a portion of Kansas City, Kansas, called *Strawberry Hill.* This area was home to many immigrants from eastern and southeastern Europe who came west in the 1800s to work in the huge packing plants that operated in the Kansas River bottoms just to the east. Many descendants of these immigrants still live in this neighborhood.

423.7   At this point I-70 curves to the southeast and crosses the Kansas River by means of the Lewis and Clark viaduct. Just to the northeast is the *confluence of the Kansas and Missouri rivers.* The intersection of the midlines of these two rivers marks the beginning point of the Missouri-Kansas boundary south of the Missouri River. The Lewis and Clark expedition camped at the mouth of the Kansas River during its exploration of the Louisiana Purchase.

424.2   *Kansas-Missouri state line.*

Figure 121—The confluence of the Kansas and Missouri rivers, just north of the Lewis and Clark viaduct at milepost 424 along I-70.

Additional Reading

The following books contain additional information about geology and Kansas. Nearly all are nontechnical and easily understood. In addition, the Kansas Geological Survey publishes a number of technical and nontechnical books and maps describing Kansas geology; a list of those publications is available on request.

Preston Cloud, *Cosmos, Earth, and Man: A Short History of the Universe* (New Haven: Yale University Press, 1978), 372 pp.

L. B. Halstead, *Evolution of the Mammals* (Naples, Italy: Eurobook), 1978.

N. Hotton III, *The Evidence of Evolution* (New York: American Heritage Publishing, 1968).

W. Ley, *Worlds of the Past* (New York: Golden Press, 1971).

John Madson, *Where the Sky Began: Land of the Tallgrass Prairie* (Boston: Houghton Mifflin, 1982), 321 pp. An introduction to the grasses and animals of the tall-grass prairie.

Grace Muilenburg and Ada Swineford, *Land of the Post Rock: Its Origins, History, and People* (Lawrence: University Press of Kansas, 1975), 207 pp. A description of the geology and uses of post rock limestone in north-central Kansas.

Brian O'Neill, *Kansas Rock Art* (Topeka: Kansas State Historical Society, 1981), 34 pp. This well-illustrated book describes the Indian petroglyphs found throughout the state.

Frank Press and Raymond Siever, *Earth,* 2d ed. (San Francisco: W. H. Freeman and Co., 1978), 649 pp. An introductory earth science textbook.

John Rydjord, *Kansas Place-Names* (Norman: University of Oklahoma Press, 1972), 613 pp. This book provides the source of names for many cities, rivers, and other locations in Kansas.

Donald E. Trimble, *The Geologic Story of the Great Plains,* U.S.G.S. Bulletin 1493 (Washington, D.C.: U.S. Geological Survey, 1980), 55 pp.

Glossary

The words in italics may be found defined in this glossary in alphabetical order.

ALGAE—A class of primitive, single-celled plants and common seaweeds that contain chlorophyll. Some forms are important in the formation of *carbonate rock*.

ALLUVIUM—The unconsolidated sediment—composed of sand, *gravel*, or *clay*—that has been deposited by water, including rivers and lakes. The alluvium along many Kansas rivers often acts as an *aquifer*, providing a water source.

ANOMALY—A departure from the norm. In *geophysics*, a magnetic anomaly occurs in an area where magnetic levels are significantly higher or lower than expected. The Big Springs magnetic anomaly in eastern Kansas, for example, is an area of high magnetic levels.

ANTHRACITE—A hard, glassy *coal* that contains 92 to 98 percent carbon. There is virtually no anthracite coal in Kansas.

AQUIFER—A formation that is capable of holding and yielding significant amounts of *ground water*. In most aquifers, ground water is held in the pore spaces between grains of rock. The *Ogallala* aquifer, which underlies the western third of Kansas, is composed mostly of sand and *gravel*.

BASEMENT—Mostly *metamorphic* and *igneous* rocks that underlie *sedimentary* rocks, much the way a basement underlies a house. Kansas basement rock is 600 to 8,700 feet below the surface, and nearly all of it is Precambrian in age.

BENTONITE—Composed largely of *clay* minerals, this *rock* will absorb water and swell when wet. Because of this property, it is used extensively as a sealant during drilling of oil, gas, and water wells.

BITUMINOUS coal—The soft, most common grade of *coal*. All coal mined in Kansas is bituminous.

BLOWOUT—A shallow circular depression formed by wind *erosion*. Common in western Kansas.

BRACHIOPOD—A shelled marine *invertebrate* that is a common *fossil* in much of eastern Kansas.

CALCITE—The mineral calcium carbonate. It is the principal component of *limestone* and one of the most common minerals in Kansas.

CALICHE—Common in western Kansas, this is a *rock* that is cemented by calcium carbonate. Formed during the *Tertiary* period, caliche is whitish in color.

197

CARBONATE ROCK—Rocks composed of carbonate minerals, such as *calcite*. Common Kansas carbonate rocks are *limestone* and *dolomite*.

CHALK—A variety of *limestone* that is composed of the skeletons of microscopic, ocean-dwelling plants and animals. *Cretaceous*-age chalk formations are found in western Kansas, and chalk is sometimes referred to as the state rock of Kansas.

CHERT—Also called flint, this is a fine-grained, noncrystalline *sedimentary rock* composed of silica. Layers of chert are common in the *limestone* of eastern Kansas. Because it is hard, chert has helped make the Flint Hills resistant to *erosion*.

CLASTIC—Clastic rocks are composed of fragments of *rock* that have been transported and redeposited. *Sandstones* are an example of a clastic rock.

CLAY—Material made up of particles smaller than 1/256 millimeter, so small they can only be seen with a microscope. These particles were formed by the *weathering* of solid *rocks* and were carried to some body of water, such as a lake or stream, where they settled to the bottom. Clay deposits are common throughout Kansas.

CLEAVAGE—In certain *rocks,* cleavage is the tendency to split along parallel planes.

COAL—*Rock* that contains more than 50 percent carbon (by weight), formed from plant remains by heat and pressure caused by the weight of younger, overlying sediments. Kansas coal is either *bituminous* or *lignite.*

CONCRETION—An accumulation of mineral or *rock* that is, or was at one time, surrounded by other softer *rock* material. As the softer material weathers away, the concretion remains; most are spherical in shape. The round sandstone shapes at Mushroom Rocks State Park are examples of concretions.

CONE-IN-CONE—Although it may look like a fossil, this is a sedimentary structure that has a scaly appearance and resembles a cone within a cone. It is easily mistaken for petrified wood.

CONGLOMERATE—*Sedimentary rock* composed of smooth, rounded pebbles, *gravel,* or boulders that have been cemented together.

CONTINENTAL DRIFT—The movement of continents. Geologists theorize that continents are a part of plates that drift across the surface of the earth, much the way an iceberg drifts through water. They believe that continents were once part of one huge land mass, before they broke up and drifted to their present positions.

CORALS—A group of bottom-dwelling, marine *invertebrates* that may grow as individuals or in colonies, sometimes creating reefs. They create external skeletons that are common *fossils* in Kansas *rocks.*

CRETACEOUS—The period from approximately 138 to 63 million years ago. Parts of western Kansas were covered by a shallow sea that was home to a variety of fish, birds, and reptiles that were the source of a number of spectacular *fossils.*

CRINOIDS—A marine *invertebrate* with a cup-shaped head attached to a jointed stalk. Also known as sea lilies, their *fossils* are common in eastern Kansas, particularly the disc-shaped pieces of their stem.

CROSS-BEDDING—Inclined rock layers that have been deposited at an angle, when compared to other layers of rock. Cross-bedding occurs in *sedimentary rock* deposited by wind or water; it is apparent in the sandstone concretions at Mushroom Rocks State Park and at Rock City.

CRUST—The outermost layer of the earth; it is several miles thick.

CRYSTAL—The regular arrangement of molecules or atoms in a *mineral*. Sometimes in large crystals these molecular arrangements are visible to the naked eye.

CUESTA—A ridge with a steep face at one end and a gently sloping face at the other. This topography is common in the Osage Cuestas area of southeastern Kansas.

CYCLOTHEM—A cycle of deposition that probably occurred in shallow seas and low-lying land areas. As the shoreline advanced and withdrew, series of beds were deposited in a predictable sequence: *sandstone, shale, limestone,* shale, sandstone. Cyclothemic deposition probably occurred in eastern Kansas during the *Pennsylvanian* period.

DELTA—The body of sediments deposited at the mouth of a river.

DENTICLES—A small tooth or projection. Also, small toothlike structures in shark skin.

DEPOSITION—The accumulation of sediments.

DOLOMITE—A *mineral* composed of a mixture of equal parts of *carbonate* of calcium and magnesium. This is also the name for *rock* composed largely of the *mineral* dolomite. Dolomite resembles *limestone*. It is common in the subsurface in Kansas and is quarried in the central part of the state.

DRAW—A small valley or ravine; usually a natural drainage-way.

EARTHQUAKE—Sudden movement of the earth resulting from the abrupt release of slowly building strain. Kansans have recorded a number of small earthquakes during the state's history, although none were larger than about 5.5 on the Richter Scale.

EROSION—The process in which *rock* and soil are loosened, broken down, and transported.

ERRATIC—A *rock* fragment, usually large, that has been transported from far away. The *quartzite* boulders found in northeastern Kansas, transported by glaciers, are called glacial erratics.

ESCARPMENT—A steep slope or cliff.

EVAPORITES—Sediments that were deposited as a result of evaporation. *Gypsum* and salt are two common types of evaporites, deposited when a shallow sea evaporated in Kansas during the *Permian* period.

EXTRUSIVE ROCK—*Rock* from lava or other material that has been spewed out on the earth's surface.

FAULT—A fracture in the earth's crust where two sides have moved relative to each other. Kansas has a number of small faults; a long fault zone, extending across the eastern third of the state, occasionally produces small earthquakes.

FORMATION—A layer of rock that is, or was once, horizontally continuous. In *stratigraphy,* a formation is the basic unit used to describe rocks; formations can be lumped together into groups, or subdivided into members.

FOSSIL—The outline, traces, or body part of a plant or animal that has been preserved in rock. This includes not only the remains of animals, but other evidence such as tracks.

FRACTURE—In mineralogy, this is the way a *mineral* looks when broken. Some *rocks,* for instance, have conchoidal (or shell-like) fracture. In rocks, fractures are breaks caused by folding or faulting.

FUSULINID—An important group of extinct, one-celled ocean-dwellers. They were shaped like a grain of wheat, and their fossils are common in the *Pennsylvanian* and *Permian* rocks of eastern Kansas.

GALENA—A *mineral* that is the principal ore of lead. It can be found in southeastern Kansas, where it was mined from about 1870 to 1970.

GEODE—Round, hollow bodies that measure from one to twelve inches in diameter. The inside is filled with *crystals,* usually *quartz,* that point inward. They are most commonly found in *limestone,* less often in *shale.*

GEOLOGY—The science that studies the origin, structure, composition, and history of the earth.

GEOPHYSICS—The branch of physics dealing with the earth, including seismology, gravity, magnetics, and other disciplines. In Kansas, geophysics is often applied to the search for oil, gas, and *ground water.*

GLACIAL TILL—Sediment that has been carried or deposited by glaciers. Much of northeastern Kansas is blanketed by a layer of glacial till that includes a large component of *clay.*

GRANITE—Often used to refer to any light-colored, coarse-grained *igneous rock.* With the exception of a few small areas—and *rocks* transported into Kansas—there is virtually no granite at the surface in Kansas. It does make up part of the *basement* rock underlying Kansas.

GRAVEL—The accumulation of small pebbles that have been rounded by water. As the word is commonly used, the pebbles are generally not much larger than a chicken egg.

GROUND WATER—Underground water that is found in the pore space of rocks. These water-saturated formations are called *aquifers.*

GYPSUM—A soft mineral that is common in *sedimentary rock.* It is one of the first minerals formed when sea water evaporates. Layers of gypsum are found in rocks of *Permian* age in south-central Kansas. It is mined for use in plaster.

HARDNESS—A *rock* or *mineral*'s resistance to scratching. It is an important characteristic used to identify specimens.

HOGBACK—Similar to a *cuesta,* this is a ridge that ends in two steep, equal slopes.

HUMBOLDT FAULT ZONE—A series of earthquake-producing *faults* that run from Nemaha County in northern Kansas to Sumner County in southern Kansas. The zone is named for Humboldt, Nebraska, and not the town of the same name in Kansas.

IGNEOUS ROCK—One of the three types of *rock,* this is formed by the solidification of molten rock or magma. It occurs both at the surface and deep beneath the earth.

INORGANIC—Compounds not containing carbon.

INTRUSIVE ROCK—*Rock* that consolidated from magma below the ground, without reaching the surface. Much of the *basement* rock in Kansas is intrusive rock.

INVERTEBRATE—An animal without a backbone or spinal column.

KIMBERLITE—A type of *igneous rock* found in volcanic pipes that have forced their way to the surface from deep underground. Kimberlite often contains garnets and is the only known source of diamonds. In Kansas, a series of kimberlites is found in Riley County.

LIGNITE—A brownish-black, soft coal that has about 70 percent carbon content. Lignite was an important fuel in Kansas in the late 1800s and early 1900s. It was mined extensively in central and northern Kansas.

LIMESTONE—One of the most common rocks in Kansas, it is a *sedimentary rock* composed largely of calcium carbonate.

LOESS—A silty, dusty sediment that has been deposited by wind. It is often rich in *clay* minerals and is found throughout Kansas, particularly in steep bluffs in the northeastern and northwestern part of the state.

LUSTER—The shine of a *mineral*'s surface, this is one of the characteristics used to describe minerals.

MASSIVE—A term used to describe rocks that have a homogenous, even structure, not banded or layered. It is also used to describe rocks that occur in thick beds, such as the *limestone* layers of eastern Kansas.

MEANDER—Broad, semicircular curves in a stream bed. Meanders are common features in the Kansas, Saline, Smoky Hill, and other rivers in the state.

MEMBER—Stratigraphers describe *rock* layers in basic units called *formations,* but formations can be further subdivided into rock layers, called members, that have similar characteristics. The Oread Formation, for example, comprises a number of smaller rock layers called members, each with its own name.

METAMORPHIC—One of the three groups of rocks, this describes *rock* that has been significantly changed by heat, pressure, or chemical processes. Metamorphic rocks are rare in Kansas, except in a small area of Woodson County.

METEORITE—A mass of *mineral* or *rock* matter that falls from outer space to the earth's surface. Meteors, on the other hand, are similar but burn up before they reach the earth's surface.

MICA—Any member of a group of minerals that are common in Kansas. Mica often has a sheetlike *cleavage.*

MINERAL—A naturally formed, solid, inorganic element or compound that has a definite composition.

MISSISSIPPIAN—The period of geologic history from about 360 million to 330 million years ago. Rocks deposited during the Mississippian are found in Kansas only in the extreme southeastern corner of the state.

MOHS HARDNESS SCALE—A scale that shows the hardness of minerals. The higher the number, the harder the mineral. In the scale, talc is assigned a hardness of 1, *gypsum* 2, *calcite* 3, fluorite 4, apatite 5, orthoclase 6, *quartz* 7, topaz 8, corundum 9, and diamond 10.

MOSASAUR—An extinct marine reptile that grew to a length of fifteen to twenty feet. Mosasaur fossils are relatively common in the *Cretaceous* formations of western Kansas.

NATIVE ELEMENT—Any element occurring uncombined in nature. The only native elements that occur naturally in Kansas are sulfur and the gas helium.

OGALLALA FORMATION—A *Tertiary*-age formation that occurs in the western third of Kansas. Made up mostly of sand, *gravel,* and *silt.* The Ogallala is an important source of *ground water* for much of western Kansas, where the formation acts as *aquifer,* storing water in the rock's pore spaces.

OÖLITES—Spheres of *carbonate* material, usually about the size of a grain of sand. Oölites are formed by chemical precipitation in a warm ocean, and they can form some types of *limestone* that are common in southeastern Kansas.

ORGANIC—In chemistry, organic refers to any compound containing carbon. In general, organic refers to any material composed of or derived from living organisms.

OUTCROP—The location where a rock layer is naturally exposed at the surface.

OUTLIER—Portions of a rock formation that are detached or away from the main body. Castle Rock in western Kansas is an example of an outlier.

OXBOW LAKE—A crescent-shaped lake created when a stream abandons a river bend and changes course. The Kansas River has created an oxbow lake near Lawrence.

OXIDE MINERAL—A *mineral* formed by the combination of oxygen with an element. An example is hematite, a compound of iron and oxygen.

PALEONTOLOGY—The study of past life-forms and their evolution, based on *fossil* remains.

PENNSYLVANIAN—The period of geologic history from about 330 to 290 million years ago. Pennsylvanian-age rocks are common in eastern Kansas.

PERIDOTITE—A coarse-grained igneous *rock*. It makes up much of the kimberlite found in Riley County and the *intrusives* in Woodson County.

PERMIAN—The geologic period from about 290 to 240 million years ago. Permian-age rocks are found at the surface in Kansas in roughly the same area covered by the Flint Hills and the Red Hills.

PETROLEUM—A naturally occurring hydrocarbon liquid. Most Kansas petroleum is produced in central, western, and southeastern Kansas.

PLEISTOCENE EPOCH—The geologic term for the time from about 2 million to 10,000 years ago. Much of the *loess* and dune sand, found along several Kansas rivers, was deposited during the Pleistocene. It also includes the time when glaciers moved into northeastern Kansas.

PTEROSAURS—These are extinct flying reptiles. Their *fossils* have been found in *Cretaceous* rocks of western Kansas.

PYRITE—Iron sulfide, also called fool's gold. Commonly found in yellow cubes in several parts of Kansas.

QUARTZ—A hard *mineral* composed of *silica* and oxygen; it can form a six-sided *crystal*. It is the most common rock-forming mineral.

QUARTZITE—A *rock* formed by the metamorphism of sandstone. Red quartzite boulders, brought into Kansas via glaciers, are common in the northeastern part of the state.

RECHARGE—The replenishment of *ground water* in an *aquifer*. Recharge in western Kansas is relatively low, often less than a foot per year, while in other parts of Kansas it is much more.

ROAD-CUT—The location where a road cuts through layers of rock.

ROCK—Any naturally occurring mass of one or more *minerals*.

SANDSTONE—*Rock* that is formed by grains of sand that have been cemented together. Sandstone was generally formed near the shores of ancient seas and is found throughout central and eastern Kansas.

SEDIMENTARY ROCK—One of three major types of *rock*, this is formed by the *deposition* of sediments in water or by the wind. Characteristically deposited in layers or beds, sedimentary rocks are the most common type on the surface of Kansas and cover about 75 percent of the land's surface on the globe.

SEISMOGRAPH—An instrument for magnifying and recording movements in the earth's surface, such as *earthquakes*.

SHALE—A *sedimentary rock* made up mostly of *clay* minerals. Commonly breaks in sheets or along parallel planes. Common throughout eastern and central Kansas.

SILICA—Silicon dioxide, silica forms such crystalline minerals as *quartz*.

SILICATES—Minerals containing an element, silicon, and oxygen. In Kansas common silicates are *mica* and *quartz.*

SILICIOUS—Containing *silica,* or abundant in *quartz.*

SILT—Sediment composed of small particles between $1/16$ and $1/256$ millimeter in diameter. Silt is commonly deposited by rivers in Kansas and is a principal component of *loess,* a wind-deposited silt.

SINKHOLES—A funnel-shaped depression in the land's surface caused by the slow, or sudden, collapse of the roof of a subterranean cavern. Common in central and southern Kansas in areas underlain by salt or gypsum; and in southeastern Kansas where the ground has been undermined for lead, zinc, or *coal.* Some are as small as a few feet in diameter; others, such as the Big Basin in Clark County, are a mile across.

STRATA—Layers of *rock* with similar characteristics.

STRATIGRAPHY—The study of layers of *sedimentary rock,* including *deposition,* age, distribution, and other characteristics.

STREAK—The color of mineral dust remaining when a *mineral* specimen is rubbed over a piece of unglazed porcelain, also called a streak plate. It is another characteristic used to identify minerals.

SULFATES—Minerals that consist of an element combined with sulfur and oxygen. Examples found in Kansas include *gypsum* and anhydrite.

SULFIDES—Minerals formed by the combination of an element with sulfur. Many are ore-producing minerals, such as *galena* and sphalerite, and in Kansas most sulfides are found in the southeastern corner of the state.

TERTIARY—The period of geologic history from about a million to 70 million years ago.

TOPOGRAPHY—The shape of the earth's surface. Kansas topography, for example, consists of rolling hills in much of the central and eastern parts of the state and plains in much of western Kansas.

TRILOBITE—A group of extinct marine *invertebrate* animals that had a jointed outer skeleton. Their fossils are sometimes found in *Pennsylvanian*-age rocks in eastern Kansas.

VERTEBRATE—An animal with a backbone or spinal column.

VOLCANIC ASH—The fine-grained material, usually glass, that is thrown out during a volcanic eruption. By definition, ash is composed of particles smaller than four millimeters in size. Ash deposits found throughout central and western Kansas are the result of volcanoes in New Mexico, Wyoming, and California that spewed out the ash, which blanketed Kansas and then collected in streams and lakes.

VUG—A cavity.

WATER TABLE—The upper level of an area where an *aquifer* is saturated with water. Also, the level at which a well fills with water.

WEATHERING—The chemical and physical processes that decay and break up *rock.*

Index